T0310305

Design Optimization of Fluid Machinery

Design Optimization of Fluid Machinery

Applying Computational Fluid Dynamics and
Numerical Optimization

Kwang-Yong Kim
Inha University
Incheon
Republic of Korea

Abdus Samad
Indian Institute of Technology Madras
Chennai
India

Ernesto Benini
University of Padova
Italy

Registered Offices
John Wiley & Sons, Inc., 111 River Street, Hoboken, NJ 07030, USA
John Wiley & Sons Ltd, The Atrium, Southern Gate, Chichester, West Sussex, PO19 8SQ, UK

Editorial Office
The Atrium, Southern Gate, Chichester, West Sussex, PO19 8SQ, UK

For details of our global editorial offices, customer services, and more information about Wiley products visit us at www.wiley.com.

Wiley also publishes its books in a variety of electronic formats and by print-on-demand. Some content that appears in standard print versions of this book may not be available in other formats.

MATLAB® is a trademark of The MathWorks, Inc. and is used with permission. The MathWorks does not warrant the accuracy of the text or exercises in this book. This work's use or discussion of MATLAB® software or related products does not constitute endorsement or sponsorship by The MathWorks of a particular pedagogical approach or particular use of the MATLAB® software.

Library of Congress Cataloging-in-Publication Data

Names: Kim, Kwang-Yong, 1956- author.
Title: Design optimization of fluid machinery : applying computational fluid
    dynamics and numerical optimization / Kwang-Yong Kim, Professor, Inha
    University, Incheon, Abdus Samad, Associate Professor, Indian Institute of
    Technology Madras, Chennai, India, Ernesto Benini, Professor, University
    of Padova, Italy.
Description: Hoboken, NJ : Wiley, 2019. | Includes bibliographical references
    and index. |
Identifiers: LCCN 2018044844 (print) | LCCN 2018045697 (ebook) | ISBN
    9781119188322 (Adobe PDF) | ISBN 9781119188308 (ePub) | ISBN 9781119188292
    (hardcover)
Subjects: LCSH: Computational fluid dynamics.
Classification: LCC TA357.5.D37 (ebook) | LCC TA357.5.D37 K56 2019 (print) |
    DDC 620.1/064–dc23
LC record available at https://lccn.loc.gov/2018044844

Cover Design by Wiley
Cover Image: Courtesy of Kwang-Yong Kim

Set in 10/12pt WarnockPro by SPi Global, Chennai, India

Printed in Singapore by C.O.S. Printers Pte Ltd

10 9 8 7 6 5 4 3 2 1

Chihee, Minji, and Soonwook
– Kim

My wife Husnahara, son Sohail and daughter Arshi
– Samad

My beloved family
– Benini

Contents

Preface

This book introduces methods for design optimization and their applications to design of fluid machinery, such as pumps, compressors, turbines, fans, and so on. Although flow analysis in a complex flow passage is difficult and takes a lot of computing time unlike structural analysis, design optimization based on three-dimensional flow analysis has become popular even in the fluid machinery area in the last couple of decades with recent developments in computing power. Design technology of fluid machinery has developed with the development of fluid mechanics over a long time. Thus, before computational fluid dynamics (CFD) became practical, there were various design methods using empirical formulas and approximate analysis. Now, fluid machinery design has been further improved with the application of design optimization based on CFD as an additional design procedure.

Inverse design methods, where the optimum geometry of a fluid machine is deduced from prescribed objectives, require low computational cost but it is difficult to specify the target flow field. Thus, design optimization, where optimum objectives are found by changing the design variables, has recently become popular in fluid machinery design. This book is concerned with the design optimization method. The design optimization methods can be classified into gradient-based and statistical methods. Because the computing time depends on the number of design variables, gradient-based methods are not suitable for design problems that have a large number of design variables, except for the adjoint method. As a statistical approach, surrogate-based optimization methods are widely used in the design optimization of turbomachinery due to their easy implementation and affordable computing time. Surrogate modeling of objective function(s) largely reduces the number of objective function evaluations required for optimization, and thus is suitable for fluid machinery design where CFD analysis takes a long computing time. This book introduces general methods of surrogate-based optimization and their applications to fluid machinery.

Design objectives, such as efficiency, pressure ratio, weight, and so on, and geometrical/operational design variables are set depending on the characteristics of the fluid machinery to be optimized. From the huge number of examples of design optimizations for different kinds of fluid machinery presented in this book, fluid machinery designers are expected to have some idea as to how the optimization methods, design objective(s), and variables are selected in order to achieve their design goals.

This book aims to provide engineers and graduate students in universities with a general understanding of surrogate-based design optimization of fluid machinery using two- or three-dimensional numerical analysis of fluid flow, and also to introduce applications of various design optimization techniques to different types of fluid machinery.

The authors are grateful to the following graduate students for their assistance in completing this book: Tapas, Karthikeyan, Ezhil, Madhan, Hamid, Murshid, and Paresh at IIT Madras, and Hyeon-Seok Shim, Sang-Bum Ma, Jun-Hee Kim, and Han-Sol Jeong at Inha University.

Kwang-Yong Kim
Abdus Samad
Ernesto Benini

1

Introduction

1.1 Introduction

Fluid machinery is classified as those devices that transform fluid energy to shaft work or vice versa. The history of fluid machinery is long and the design technology of fluid machinery has developed with the development of fluid mechanics. Although the exact governing equations for single phase Newtonian viscous fluid, that is, the Navier–Stokes equations, were derived in middle of the nineteenth century, various approximate analysis methods, such as those with inviscid assumptions, were still used in the analysis of fluid flow before the Navier–Stokes equations were practically solved by numerical analysis using electronic computers more than a hundred years later. Thereafter, owing to the rapid development of computers, computational fluid dynamics (CFD), which solves the governing differential equations, becomes practical in the analysis of fluid flow.

Due to the complexity of the flow path in fluid machinery, application of three-dimensional (3D) CFD to the aerodynamic or hydrodynamic analysis of fluid machinery was somewhat delayed, but recently, CFD has been widely used in the analysis and design of fluid machinery. In the early stages, CFD was only used in the analysis of flow fields in fluid machinery due to the long computing time. But, continuous enhancement in computing power made the design optimization of fluid machinery using CFD practical. Thus, now CFD is utilized not only in the analysis of the flow in fluid machinery, but also in design through systematic optimization algorithms. However, instead of replacing the conventional design methods of fluid machinery, design optimization using CFD is being used as a supplementary design due to excessive computing times when it is used for the entire design of a fluid machine.

A typical design procedure recommended for the design of fluid machinery using CFD is as follows; a preliminary design using an approximate analysis method to determine a basic model of the fluid machine considered, a parametric study using 3D CFD to find the sensitivities of performance parameters on some selected geometric/operational parameters, and single- or multi-objective design optimization of the fluid machine using the design variables selected through parametric study. The design optimization requires repeated evaluations of the objective function(s), which is selected among the performance parameters of the fluid machine, and the number of objective function evaluations depends on the number of design variables and the optimization algorithm employed. An increase in the number of design variables in an optimization is generally expected to improve the results of the optimization, but the number of design variables for optimization is restricted mainly by the computational time. Therefore, design

Design Optimization of Fluid Machinery: Applying Computational Fluid Dynamics and Numerical Optimization,
First Edition. Kwang-Yong Kim, Abdus Samad, and Ernesto Benini.
© 2019 John Wiley & Sons Singapore Pte. Ltd. Published 2019 by John Wiley & Sons Singapore Pte. Ltd.

optimization could become more popular in fluid machinery design if computing power is further enhanced.

1.2 Fluid Machinery: Classification and Characteristics

The fluid machines that transform fluid energy to shaft work are called turbines; more specifically, gas, steam, wind, and hydraulic turbines, depending on the working fluid. The other group of fluid machines that transform shaft work to fluid energy includes pumps, fans, blowers, and compressors. All the machines in this group using liquids are called pumps. But, if gases are used for the work, machines in this group are divided into fans, blowers, and compressors, depending on the magnitude of pressure rise.

Fluid machinery is also divided into two categories; turbomachinery and positive displacement fluid machinery. In turbomachinery, rotating blades (rotors) perform continuous energy transfer from or to the fluid flow passing through the blade passages. However, in positive displacement fluid machinery, there is a displacement of a certain amount of working fluid without relative motion between the fluid and moving part of the machine in rotating or reciprocating motion. In other words, the working fluid does not flow in certain parts of these machines. The following sections in this chapter are mostly concerned with turbomachinery.

Turbomachinery can be also categorized according to the change in the flow direction through the impeller as shown in Figure 1.1. If the flow direction does not change through the impeller, those machines are called axial flow turbomachines. Machines where the flow direction changes perpendicularly through the impeller are called radial flow (or centrifugal) turbomachines. If the change in flow direction is neither axial nor radial, the machines are called mixed flow turbomachines. Also, the rotors of

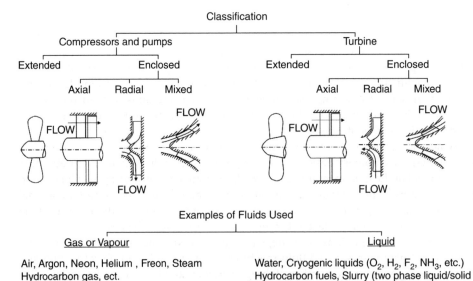

Figure 1.1 Classification of turbomachinery types. Source: Reprinted from Lakshminarayana 1996 (Figure 1.1 from original source), © 1996, with the permission of John Wiley & Sons, Inc.

turbomachinery may be enclosed in a casing or exposed to the environment without. Most turbomachines belong to the former group of enclosed turbomachines, but some, such as the wind turbine, prop fan, and ship propeller, belong to the latter group of extended turbomachines.

An important flow phenomenon found only in fluid machinery employing liquid as the working fluid is cavitation, which indicates generation of gas bubbles at normal temperature of operation due to a decrease in the local static pressure. In pumps or hydraulic turbines, cavitation occurs by the rotating blades that cause low local pressure. Repeated breaking down of bubbles near the solid wall induces erosion damage and also noise. Thus, cavitation is an important factor to be considered in the design of hydraulic machinery. On the other hand, in fluid machines that use gas as a working fluid and operate at high speed, the compressibility of gas causes unique flow phenomena such as shock waves that are not found in hydraulic machinery.

A typical parameter, which is used to classify various types of turbomachinery, is specific speed. The specific speed is defined as a non-dimensional parameter combining operating parameters of turbomachinery as follows;

$$N_s = NQ^{1/2}/(g\Delta H)^{3/4} \tag{1.1}$$

Constant specific speed indicates the flow conditions that are similar in geometrically similar turbomachinery. However, if the gravitational acceleration, g, is assumed constant, the parameter becomes a dimensional parameter, $NQ^{1/2}/(\Delta H)^{3/4}$. The specific speed, N_s is the most important parameter in turbomachinery that can be used in the selection of turbomachinery type as shown in Figure 1.1. The range of specific speeds for a specified type of turbomachinery shown in Figure 1.2 indicates the range where the turbomachine type shows maximum efficiency.

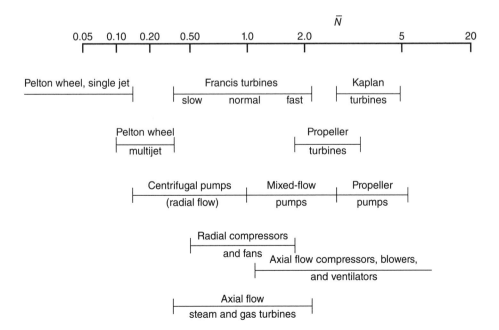

Figure 1.2 Specific speed suitability ranges of various designs. Source: Csanady 1964.

1.3 Analysis of Fluid Machinery

Analysis of turbomachinery should involve the analyses in a variety of fields; fluid mechanics, thermodynamics, solid mechanics, rotor dynamics, acoustics, material science, mechanical control, manufacturing, and so on. However, aerodynamic/hydrodynamic performance is essential in the evaluation of the basic performance of turbomachinery. Since it is difficult to include all the analyses here, only aerodynamic/hydrodynamic analysis and design methods are introduced in this chapter.

The history of turbomachinery is quite long. For example, waterwheels have been utilized by human beings for several thousands of years. The design of such ancient fluid machines was required even before the basic *theory* of fluid dynamics was set up. Therefore, the analysis method of fluid machinery was developed with the development of fluid mechanics. Until the numerical calculation of 3D Navier–Stokes equations became possible by using electronic computers in the middle of the twentieth century, analysis of turbomachinery was based on various approximate fluid mechanical theories as shown in Table 1.1. Analysis using inviscid equations and one-dimensional analysis using empirical formulas for energy losses are typical examples of such approximate analysis. Thus, many simple design methods based on these approximate analyses have developed over a long time, but the rapid development of electronic computers since the late twentieth century makes the numerical calculation of full Navier–Stokes equations practical. And, recently, 3D CFD has even become popular in the analysis of turbomachinery.

Direct numerical simulation (DNS) of Navier–Stokes equations for the wall-bounded turbulent flow was first realized by Kim et al. (1987). However, DNS cannot be used in analyzing practical flows due to excessive computational expenses. Since the numbers of spatial meshes and time steps required for DNS increase rapidly as the Reynolds

Table 1.1 Various approximations for flow analysis.

Governing equations	Assumptions
Stream function equation	Two-dimensional (2D) potential flow
Laplace equation (stream function or velocity potential)	Irrotational inviscid flow
Euler equations	Inviscid flows
Boundary layer equations	Boundary layer approximations
Stream function and vorticity equations	2D viscous flows
Parabolized Navier–Stokes (PNS) equations	If the streamwise pressure gradient can be prescribed in thin-layer Navier–Stokes (TLNS) equations, the numerical solution is independent of downstream boundary conditions
TLNS equations	If thickness of boundary layer is smaller than the body length, the streamwise diffusion terms can be neglected in Navier–Stokes equations
Full Navier–Stokes equations	

number increases, DNS analysis of turbomachinery flows is still impractical. Large eddy simulation (LES), which solves equations only for large eddies of turbulence by modeling small eddy motion, is an approximation of DNS, but LES still needs a huge amount of computing time and storage for turbomachinery analysis, as in the recent work of Pacot et al. (2016). Therefore, analysis using Reynolds-averaged Navier–Stokes (RANS) equations is the only practical method to solve full Navier–Stokes equations for turbulent flows in turbomachinery, and is thus implemented in most commercial CFD software. Because RANS equations are obtained by using Reynolds decomposition of instantaneous quantities, a turbulence closure model must be used for the Reynolds stress components to close the problem. However, no single turbulence closure model (Wilcox 1993) developed so far guarantees sufficiently accurate solutions for all types of turbulent flows. As turbulence models, the two-equation models, k-ε (Launder and Sharma 1974), k-ω (Wilcox 1988), and shear stress transport (SST) (Menter 1994) models have been most widely used for practical calculations. The SST model combines k-ε and k-ω models by implementing the k-ω model in the near-wall region and k-ε model in the region far from the wall.

The flow analysis methods for turbomachinery were classified by Lakshminarayana (1996) as shown in Figure 1.4. The flow in turbomachinery is complicated and 3D as shown in Figure 1.3. And, thus, full Navier–Stokes equations are required to be solved to resolve the complex viscous flow structures including flow separation. However, the zonal method is used when the solution of full Navier–Stokes equations in the computational domain is expensive. In this method, multiple zones are defined in the computational domain, different approximations are applied to different zones, and the solutions are integrated into the whole domain to get a complete solution. This method

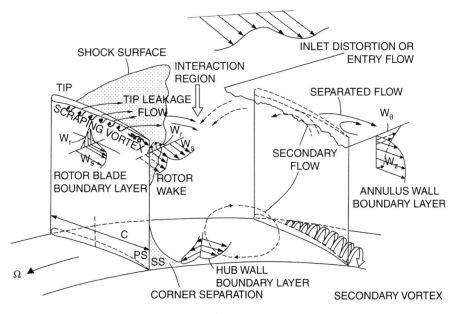

Figure 1.3 Flow structure in a rotor passage of an axial flow compressor. Source: Reprinted from Lakshminarayana 1996 (Figure 1.15 from original source), © 1996, with the permission of John Wiley & Sons, Inc.

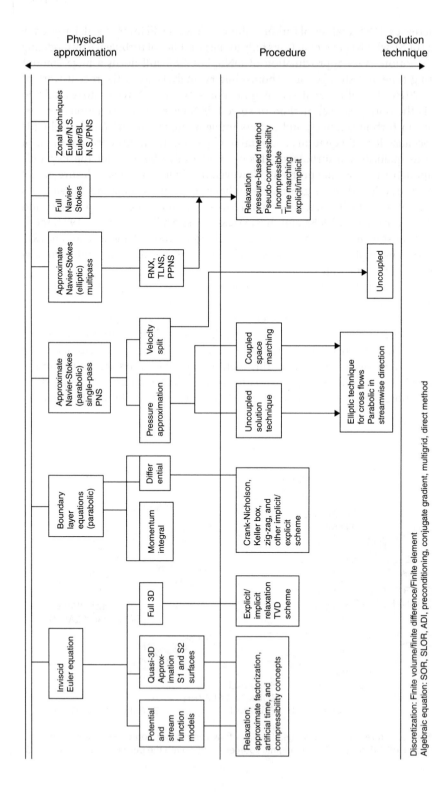

Figure 1.4 Flow analysis methods for turbomachinery. Source: Reprinted from Lakshminarayana 1996 (Table 5.2 from original source), © 1996, with the permission of John Wiley & Sons, Inc.

Discretization: Finite volume/finite difference/Finite element
Algebraic equation: SOR, SLOR, ADI, preconditioning, conjugate gradient, multigrid, direct method

is complicated but less expensive without a great loss of accuracy. The computational errors involved in the analysis arise from different sources: incomplete physical models such as turbulence closure, discretization of the governing differential equations, and the solution procedure of algebraic equations.

Although the 3D analysis of turbomachinery flow using Navier–Stokes equations has become practical, it is still impractical to perform a whole design process using design optimization based on this analysis method due to excessive computing time. Thus, for a new design of a turbomachine, a preliminary design using approximate analysis methods is still needed to determine the values of the numerous (geometrical and operational) design parameters of the machine. As a second step, through a parametric study using 3D CFD with selected design parameters, some design variables that sensitively affect the performance of the turbomachine can be determined among the tested parameters. Then, a design optimization using these design variables would further improve the performance of the turbomachine. This is a most effective way to design turbomachinery with limited computational resources because design optimization using systematic optimization algorithms requires repeated analyses of turbomachinery flow and the number of repeated analyses is roughly proportional to the third power of a number of design variables.

1.4   Design of Fluid Machinery

Complex turbomachines, such as a gas turbine engine that consists of a multistage compressor, combustor, and multistage turbines, require many engineering considerations in their design, including thermodynamic, aerodynamic, and thermal analyses. However, for most other simple turbomachines, such as fans, compressors, pumps, and turbines, a relatively simple design process has been applied. Designs of turbomachinery have been developed over a long time, along with the development of aerodynamic/hydrodynamic analysis technology. Therefore, the procedure of turbomachinery design generally consists of several steps that perform different levels of performance analysis and design.

A typical method for aerodynamic design of turbomachinery blades follows the following steps.

1.4.1   Design Requirements

Design requirements and operating conditions of a turbomachine need to be specified in terms of flow capacity, RPM (revolutions per minute), pressure rise, efficiency, noise level, and inlet flow conditions.

1.4.2   Determination of Meanline Parameters

Design parameters such as hub-to-tip ratio, tip diameter, pitch, and chord length, are determined at the meanline as representative dimensions and on the basis of specific speed charts.

1.4.3   Meanline Analysis

Meanline performance analysis is performed using thermodynamic equations, flow deviation, and pressure loss models to estimate roughly the effects of design variables on aerodynamic performance. From this parametric study, proper fan design variables with feasible ranges for high-performance design can be determined.

1.4.4   3D Blade Design

The 3D shape of the turbomachinery blades is defined using the methods proved for camber line, blade thickness distribution, and stacking line. With specified design requirements, the baseline design of blade cross-section is obtained by mean camber line. Also, the thickness distribution of the blade cross-section is built using a distribution of points defined as a fraction of camber line length. The design of the 3D blade is completed by determining the stacking line of blade elements along the blade span considering sweep and lean and by performing a conformal mapping of the planar surfaces of the blade sections to the cylindrical surfaces.

1.4.5   Quasi 3D Through-Flow Analysis

Based on the 3D blade design, a quasi 3D through-flow method analyzes the aerodynamic performance of the turbomachine using Euler's equation, pressure loss models, and the equation of motion for radial equilibrium. This analysis predicts blade-to-blade and spanwise flow distributions and provides the aerodynamic performance through the mass-averaging of predicted flow field data. However, this analysis method has a problem in predicting 3D flow structures including leakage flow, secondary vortex, and endwall boundary layer.

1.4.6   Full 3D Flow Analysis

To precisely analyze the 3D flow field and aerodynamic performance of turbomachinery, full 3D flow analysis using Navier–Stokes equations can be used. This analysis method requires complicated grid generation and a reliable turbulence closure model for the complex 3D flow field in the turbomachine, and thus much more computing time and effort than the approximated through-flow analysis.

1.4.7   Design Optimization

Owing to the recent development of numerical methods and computers, full 3D CFD analysis can be directly used for single- or multi-objective design optimization of turbomachinery. However, a design method using an optimization algorithm requires repeated evaluations of the objective function(s) using 3D CFD, which generally takes a lot of computing time and thus there is a limitation in the number of design variables for optimization. Therefore, in the initial stage of the optimization, a parametric study using a number of geometric and/or operating parameters is usually performed in order to select the design variables and their design ranges for optimization.

1.5   Design Optimization of Turbomachinery

Although analyses of complex turbulent flows in turbomachinery take a long computational time, the recent development of high-speed computers has made it practical to optimize the aerodynamic or hydrodynamic design of turbomachinery using governing equations for 3D viscous flows, such as RANS equations. Systematic optimization using high-fidelity analysis produces high-performance and reduces computational and experimental expenses in turbomachinery design.

General objectives of turbomachinery design are efficiency, pressure ratio, weight, and so on, and geometrical/operational parameters are generally used as design variables for optimization. In a design called inverse design, the optimum turbomachinery geometry is deduced from prescribed ideal flow conditions (and thus from prescribed objectives). This inverse design only requires low computational cost, but there is a difficulty in specifying the target flow field where the designer's insight and experience are required. If optimum objectives are found by changing the design variables, the design is called direct design or design optimization. The present book is mostly concerned with this design method. The design optimization methods can be classified into two categories: gradient-based and statistical methods.

The gradient-based methods are categorized into finite difference, linearized, and adjoint methods depending on how the gradients of the objective function are calculated. Because the computing time for the finite difference and the linearized methods depends on the number of design variables, these methods are not suitable for design problems with a large number of design variables. The adjoint method has an advantage in computing time because its computing time does not depend on the number of design variables; however, this method is not being widely used because of its complexity and counter-intuitive natures (Wang and He 2008).

As a statistical approach, surrogate-based optimization methods are widely used in the design optimization of turbomachinery due to their easy implementation and affordable computing time. By employing surrogate model(s) of the objective function(s), it is possible to largely reduce the number of objective function calculations required for the optimization. The modeling fidelity is important in surrogate modeling. Various surrogate models have been developed so far (Queipo et al. 2005), and weighted average models have also been suggested based on global error measures (Goel et al. 2007). The simulated annealing and the genetic algorithm are also available for optimization but are known to have relatively large computing times.

Parametric geometric modeling is an essential element in design optimization of turbomachinery. To optimize the shape of a turbomachine, the geometry must be modeled. Lieber (2003) suggested that the techniques used to describe the geometries of components and flow paths are required to have sufficient generality for the accommodation of complex configurations. Transfer of information for flow-path geometry to other functional groups must also be considered in the design process. In many turbomachinery optimizations, the Bézier curves are used to parameterize the geometry, and related control points can be used as the design variables. The B-spline curve, which is a piecewise collection of Bézier curves, is used when a single Bézier curve cannot be used for the shape due to complexity. The parameterization of turbomachinery blades by Bézier curves have two advantages: the curves can be controlled by a small number of points to produce a smooth profile and, thus, require a small number of design variables.

References

Csanady, G.T. (1964). *Theory of Turbomachines*. New York: McGraw-Hill.

Goel, T., Haftka, R., Shyy, W., and Queipo, N. (2007). Ensemble of surrogates. *Structural and Multidisciplinary Optimization* 33 (3): 199–216.

Kim, J., Moin, P., and Moser, R.D. (1987). Turbulence statistics in fully-developed channel flow at low Reynolds number. *Journal of Fluid Mechanics* 177: 133–166.

Lakshminarayana, B. (1996). *Fluid Dynamics and Heat Transfer of Turbomachinery*. New York: Wiley.

Launder, B.E. and Sharma, B., I. (1974). Application of the energy-dissipation model of turbulence to the calculation of flow near a spinning disc. *Letters in Heat and Mass Transfer* 1 (2): 131–137.

Lieber, L. (2003). Fluid dynamics of turbomachines. In: *Handbook of Turbomachinery*, 2e (ed. E. Logan and R. Roy). CRC Press.

Menter, F.R. (1994). Two-equation eddy viscosity turbulence models for engineering applications. *AIAA Journal* 32: 1598–1605.

Pacot, O., Kato, C., Guo, Y. et al. (2016). Large eddy simulation of the rotating stall in a pump-turbine operated in pumping mode at a part-load condition. *ASME Journal of Fluids Engineering* 138 (11): 111102.

Queipo, N.V., Haftka, R.T., Shyy, W. et al. (2005). Surrogate-based analysis and optimization. *Progress in Aerospace Sciences* 41: 1–28.

Wang, D. X., and He, L. (2008). Adjoint Aerodynamic Design Optimization for Blades in Multistage Turbomachines – part I: Methodology and Verification, *ASME Turbo Expo 2008* GT2008-50208.

Wilcox, D.C. (1988). Reassessment of the scale-determining equation for advance turbulence models. *AIAA Journal* 26 (11): 1299–1310.

Wilcox, D. C. (1993). Turbulence modeling for CFD. 5354 Palm Drive, CA: DCW Industries, Inc.

2

Fluid Mechanics and Computational Fluid Dynamics

2.1 Basic Fluid Mechanics

In our daily life, we come across three states of matter: solid, liquid, and gas. Even though they are dissimilar in many respects, liquids and gases differ from solids in their characteristics: they are fluids, lacking the ability to offer a stable resistance to a shearing force. Since the fluid motion continues under the action of a shear stress, a fluid can be defined as any substance that cannot resist a shear stress when at rest. Fluid mechanics has two parts: dynamics and kinematics. The kinematics describes the motion of the fluid without any consideration of the forces that cause fluid motion. Fluid motion where the forces are considered is called fluid dynamics. Governing equations are formulated by considering the balance of these forces.

2.1.1 Introduction

Fluids deform continuously under the action of shear stress, however small it may be. Though liquids and gases both exhibit the same behavior of fluid, they have peculiar characteristics of their own. Liquids are mostly considered to be incompressible. A given mass of liquid occupies a fixed volume, regardless of the shape or size of the container, and forms a free surface if the container volume is larger than that of the liquid.

Gases are relatively easy to compress than liquids. Their volume changes with pressure and are related to temperature change. A given mass of gas does not occupy a confined volume and will expand continuously unless restricted in a covered vessel. It will completely fill any container in which it is placed and, hence, gases do not form a free surface. Even gases are assumed mostly as incompressible at velocities much lower than speed of sound.

Fluid dynamics is defined as the study of fluids to find how they will behave under various conditions. Fluid dynamics allow us to study motion of fluids, so that their dynamics can be worked out for engineering purposes.

2.1.2 Classification of Fluid Flow

Fluid flows can be classified according to several criteria; for example, considering viscosity, compressibility, Mach number, and so on.

Design Optimization of Fluid Machinery: Applying Computational Fluid Dynamics and Numerical Optimization,
First Edition. Kwang-Yong Kim, Abdus Samad, and Ernesto Benini.

2.1.2.1 Based on Viscosity

All fluids have a natural resistance to flow called viscosity. In a fluid, the molecules feel an "attraction" toward other molecules. We call this attraction a "cohesive" force. It leads to surface tension in liquids. When placed in a container, the molecules also experience an attractive force toward the interiors of the container. This is called "adhesive" force. When fluid flows, viscosity results in a frictional force, both against the surface it is flowing on and within the fluid itself (White 2017). Viscosity makes the flow interesting and of course challenging to understand and calculate. It is viscosity that causes many of the physical features of a flow. Fluid can be classified as inviscid or viscous.

2.1.2.1.1 Viscous Flow

In viscous flow, frictional effects are significant. Viscosity is the fluid property quantified by the frictional force developed between two fluid layers moving relative to each other: for example, boundary layer flows.

2.1.2.1.2 Inviscid (Ideal) Flow

Inviscid flow is nothing but the viscous terms neglected in the governing equations. It is the main theoretical model for many fields of modern technology. Calculation results obtained within the framework of this model are widely used in designing flying vehicles, rockets, turbines, and compressors.

2.1.2.2 Based on Compressibility

A flow can be categorized based on compressibility, as compressible flow, and incompressible flow, which is decided by fluid density.

2.1.2.2.1 Incompressible Flow

If the effect of pressure on density of fluid is negligible, the flow is called an incompressible flow. When the flow is incompressible, the fluid volume fraction remains constant along the flow path. For an incompressible flow the equation of continuity simplifies to $\nabla \cdot v = 0$.

2.1.2.2.2 Compressible Flow

When a fluid moves at a speed equivalent to 0.3 times the speed of sound, density variation becomes predominant and the flow is compressible. Such flows do not occur easily in liquids, since high pressures of order 1000 atm are required to produce sonic velocities in liquids.

Water hammer and cavitation are examples of the significance of compressibility in liquid flows. Water hammer is produced by acoustic waves reflecting and propagating in a confined liquid, for example, when a valve is closed suddenly. The resulting noise can be resembled to hammering on the pipes. Cavitation occurs when vapor bubbles are generated in a liquid flow due to local loss in pressure.

2.1.2.3 Based on Flow Speed (Mach Number)

When studying rockets, spacecraft, and high speed flow systems, the speed of flow is commonly represented in terms of a dimensionless number: Mach number is given as,

$$M = \frac{V}{c} = \frac{flow\ velocity}{Speed\ of\ sound} \tag{2.1}$$

where "c" is the velocity of sound in air (about $346 \, \mathrm{m\,s^{-1}}$) at sea level at room temperature.

2.1.2.3.1 Subsonic Flow

When Mach number is less than one ($M < 1$), a subsonic condition occurs. Compressibility effects can be ignored for lowest subsonic conditions. Shock waves do not form when the Mach number stays below unity everywhere in the flow field. In engineering standards, subsonic flows for Mach numbers below 0.3 are often considered to be incompressible.

2.1.2.3.2 Transonic Flow

As Mach number of the flow increases with increase in velocity of the body, it reaches a value nearly equal to one ($0.8 < M < 1.2$) and then the flow is believed to be transonic. Compressibility effects are predominant in transonic flows. At some place on the body, the flow speed surpasses the speed of sound and shock waves might occur. Analysis of transonic flows is tedious because of mixed regions of locally subsonic and supersonic flow and the governing equations are inherently nonlinear.

2.1.2.3.3 Supersonic Flow

Supersonic conditions take place for Mach number $1.2 < M < 3$. During this condition the disturbance waves are created due to body interacting with the incoming flow. These waves does not propagate upstream of the body, as the incoming flow is much faster than them, thus no information is passed upstream about the presence of the body. These waves get accumulated near the body as a piled up disturbance. When the flow approaches the build up disturbance the flow suddenly modifies itself to account for the presence of the body. This piled up disturbance is called the shock wave. There is a sudden jump in pressure and temperature at the instance of the shock wave.

2.1.2.3.4 Hypersonic Flow

When flow velocity exceeds five times the speed of sound, it is considered to be hypersonic flow. Large rises in temperature of the fluid with molecular dissociation and in chemical effects are observed in very high flow speeds along with friction and shock-waves when Mach number is above 3 ($M > 3$).

2.1.2.4 Based on Flow Regime

Flow regimes represent typical flow patterns exhibited by fluids as they flow under differing conditions. A flow can be laminar, turbulent, or transitional in nature. This becomes a very important classification of flows and is brought out vividly by Osborne Reynolds (1842–1912). Reynolds determined that the transition from laminar to turbulent flow is usually be found at a fixed value of the ratio that bears his name, the Reynolds number. This dimensionless number is the main parameter that governs whether the flow is laminar or turbulent. It describes the ratio of inertial forces to viscous forces in a flowing fluid. The Reynolds number for a fluid flowing in a circular pipe may be calculated as follows:

$$Re = \frac{\rho VD}{\mu} \tag{2.2}$$

where Re is the Reynolds number, V is the average velocity, ρ is the fluid density, D is the pipe diameter, and μ is the dynamic viscosity of the fluid.

2.1.2.4.1 Laminar Flow

A highly organized fluid flow defined by layers of fluid is known as laminar flow. The Reynolds number is Re < 2100 for laminar flows. For a low Reynolds number, the flow is steady, smooth, viscous, or laminar and the behavior of a fluid depends mostly on its viscosity. However, laminar flow need not necessarily be steady for all cases; unsteady laminar flows are also observed.

2.1.2.4.2 Turbulent Flow

The fluid motions that generally arise with high velocities are characterized by velocity fluctuations and are called "turbulent." For turbulent flows associated with high Reynolds numbers, the flow is unsteady, random, dissipative, diffusive and has three-dimensional vorticity fluctuations. The Reynolds number is $4000 < Re$ for internal and $Re > 10^5$ for external flows.

2.1.2.4.3 Transitional Flow

The free stream perturbations causes the flow from laminar to turbulent. The intermediate region between lamina and turbulent is termed as transitional flow. For transitional flow the Reynolds number is generally between $2100 < Re < 4000$ in case of internal flows and between $5 \times 10^5 < Re < 10^7$.

2.1.2.5 Based on Number of Phases

Fluid mechanics can also be distinguished between a single phase flow and multi-phase flow (flow made more than one phase or single distinguishable material). The last boundary (as all the boundaries in fluid mechanics) is not sharp because fluid undergoes a phase change (condensation or evaporation) during the flow and shift from a single phase flow to a multi-phase flow. Furthermore, the flow that constitutes of two or more materials may be treated as a single-phase flow too (e.g., air with dust particles). One method to classify multi-phase flow is by the materials that flow. For instance, the flow of water and oil in a single pipe is a multi-phase flow. This type of flow is engineered to decrease the cost of transporting crude oil via long piping systems from the one plant to another.

The different variants of two-phase flow are classified as:

Gas–liquid flow. Involves boiling, condensation, and adiabatic flow. They are common in power and process industries, air-conditioning, cryogenic applications, and refrigeration.

Gas–solid flow. Combustion of pulverized fuel, flow in a cyclone separator and pneumatic conveying are examples of gas–solid flow.

Liquid–solid flow. This type of flow is encountered in slurry transportation, food processing, and in various biotechnological processes.

Liquid–liquid flow. This type of flow is also characterized by the existence of a deformable interface (similar to gas–liquid flow) and processes several features similar to other two-phase flow phenomena. Liquid–liquid flow is common in petroleum industries and chemical reactors.

2.1.3 One-, Two-, and Three-Dimensional Flows

A fluid flow is said to be one-, two-, or three-dimensional (1D, 2D, and 3D, respectively) flow if the flow speed fluctuates in one, two, or three primary dimensions, respectively. Classical flow of fluids comprises of a 3D geometry and fluid velocity can fluctuate in all the three dimensions. Interpreting the velocity of the flow in 3D space is taken as $\overline{U}(x, y, z)$ in rectangular coordinates and $\overline{U}(r, \theta, z)$ in cylindrical coordinates.

Nevertheless, the change of flow speed in one dimension may be minute compared to change in other dimensions and thus may be omitted due to insignificant error. For such cases, the fluid may be modeled suitably as 1D or 2D, thus reducing the complexity. It may be possible to reduce a 3D problem to a 2D one, even a 1D one at times.

2.1.3.1 One-Dimensional Flow

All the flow parameters in a 1D flow are expressed as functions of time and one space coordinate only. It is generally the distance measured lengthwise through the center-line (not essentially straight) where the fluid is flowing. For instance, the flow through a pipe is considered 1D when pressure and velocity vary along the length of the pipe and variation across the cross-section is neglected. But in reality, any flow is never a 1D flow due to the viscosity which causes the velocity to fall to zero at the solid boundaries.

2.1.3.2 Two- and Three-Dimensional Flow

In a 2D flow, all the flow parameters are expressed as functions of time and two space coordinates (say u and v) only with no variation in the w direction. For a 3D flow, the flow parameters are functions of three space coordinates and time.

2.1.4 External Fluid Flow

The flow of fluid not enclosed by any surface is called an external flow (e.g., fluid flow over a flat plate) and if the fluid is entirely restricted by the surface, then it is treated as an internal flow (e.g., flow through a pipe/duct). If the duct is filled incompletely and there is one free surface, then it is called open channel flow. External flows are limited to wakes and boundary layers in the solid surface while internal flows are controlled by viscosity.

2.1.5 The Boundary Layer

In 1905, Ludwig Prandtl first postulated that, for low viscous flows, the viscosity acting on the fluid is insignificant at all places except adjacent to wall boundaries where the zero-slip condition should be fulfilled. The boundary layer thickness moves toward zero as the viscosity approaches zero. The flow region near the wall in which the viscous effects are prevailing is termed the boundary layer. The governing fluid transport equations inside the boundary layer may be abridged due to the layer's narrowness, and the exact solutions are likely attained in most cases. Boundary layer phenomenon gives definitions for the drag and lift calculations of bodies indifferent shapes in laminar and turbulent flows. The boundary layer principle assumes that the boundary layer is thin in comparison to length scales like the local radius of curvature or surface length. Along this thin layer, the velocity changes quickly enough for viscous effects to be significant and can occur only in high Re flows. Figure 2.1 shows the boundary layer thickness, which is inflated for representation purposes.

The fluid is flowing in from the left side of the plate with a velocity U_0 (free stream) and slows down adjacent to the surface of the plate due to zero slip condition. Therefore, a boundary layer starts to take shape starting from the leading edge of the flat plate. Large shearing stresses and velocity gradients start to develop within the boundary layer as the fluid continues further downstream. The boundary layer grows with thickness (δ)

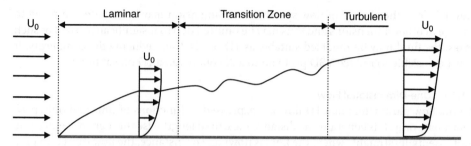

Figure 2.1 Expansion of boundary layer over a flat plate is shown along with the transition zone from laminar to turbulent boundary layer.

as more and more fluid slows down along the direction of the flow. This extends to the point where the fluid velocity reaches approximately near of the free-stream velocity. This assumption is typically made as there is no sharp line to divide the boundary layer from the free stream. The boundary layer thickness δ is comparatively smaller than any distance x from the leading edge at all times (Çengel and Cimbala 2006).

Initially, boundary layer starts out as a laminar boundary layer where the fluid particles flow in regular layers for smooth upstream boundaries. As the thickness increases at the laminar boundary layer, instability begins and transforms into a turbulent boundary layer where the fluid particles travel in a chaotic manner. Even at the turbulent boundary layer, there is still a very thin layer next to the boundary layer that is found to be in laminar motion. It is called the viscous sub-layer.

2.1.5.1 Transition from Laminar to Turbulent Flow

The development of boundary layer from laminar flow to turbulent flow as Reynolds number increases is called transition. As in boundary layer theory, laminar and turbulent boundary layers are dissimilar in many important ways. A fully turbulent boundary layer generates far more average surface shear stress compared to a laminar boundary layer.

2.2 Computational Fluid Dynamics (CFD)

CFD is a science that gives quantitative computations of fluid flow derived from the three basic conservation laws (conservation of mass, conservation of momentum, and conservation of energy) that govern fluid flow with the aid of computers. These computations are normally carried out in terms of domain geometry, the fluid flow properties, grid, and the initial boundary conditions applied to the flow domain (Anderson 1995). Modern engineers apply both CFD analysis and experimental results, which supplement each other for accurate solutions. Experimentally, global properties like lift, velocity, pressure drop, drag, or power are obtained, but CFD is used to obtain accurate flow field particulars like shear stresses, pressure, and velocity profiles or flow streamlines/streak lines (Figure 2.2).

In the past three decades, various kinds of numerical approaches have been proposed to mimic fluid flows considering a broad field of applications. These approaches

Figure 2.2 Flow streamlines through the axial turbine.

are named the finite difference method (FDM), finite element method (FEM), and finite volume method (FVM). The governing partial differential equations (PDE)s are converted into much easier algebraic equations that can be solved using computational techniques (Chung 2015). The rapid improvement in computational power in the past five decades has led to the emergence of CFD. Complex software packages have been developed that are capable of simulating transonic or turbulent flows with great accuracy and speed. CFD is practiced across various engineering domains like automotive, aerospace, metallurgy, biomedical, oceanic flows, computers and electronics, architecture, climate predictions, medical applications (heart, blood vessels), and so on.

The current state of CFD is that it can handle laminar flows effortlessly, but turbulent flows are difficult to solve without addressing the turbulence models. Different turbulence models used in CFD are described in detail later in the chapter. The Navier–Stokes equations are the fundamental basis of practically all CFD problems. The Euler equations are yielded by simplifying the Navier–Stokes equations; that is, by removing terms defining viscous actions. Further simplification produces potential equations by removing the vorticity terms. Finally, these equations can be linearized to obtain the linearized potential equations for small disturbances in subsonic and supersonic flows.

2.2.1 CFD and its Application in Turbomachinery

CFD plays an important role in analyzing fluid flow and heat transfer problems using numerical methods. Turbomachines such as fans, compressors, pumps, and turbines involve complex internal fluid flow problems that should be addressed for better efficiency. Presently, CFD is one of the most vital tools for designing and analyzing all kinds of turbomachinery.

Turbomachinery defines machines that exchange energy between a rotor and the fluid flowing through it (e.g., compressors and turbines). In a turbine, energy is transferred from the moving fluid to its rotor, while a compressor transports energy from its blades to the moving fluid. Solutions to turbomachinery problems require a strong understanding of fluid mechanics, thermodynamics, and other related subjects like compressible and incompressible gas dynamics and heat transfer. In this present era, the design of the currently available compressor, pump, or turbine is almost impossible without CFD modeling. The advantages of CFD are well understood with boosted performance, increased efficiency, and reduced weight in turbomachinery.

Microscopic and macroscopic are the two methodologies used to analyze fluid flows and the flow properties like pressure, velocity, and so on are calculated at every point in the flow domain. In the microscopic approach, the characteristic length is far greater than the molecular mean free path. In the macroscopic approach, fluid is treated as a continuum; that is, the fluid in a section is assumed to occupy every geometric point of that section. In other words, a minute volume in the fluid region whose size is zero is the limit. Flow behavior studies in turbomachines work at very low pressures, such as a turbo molecular pump that necessitates a microscopic approach. The turbomachinery flows, which are of interest in this book, are all in continuum and therefore the macroscopic approach is the most suitable methodology.

2.2.1.1 Advantages of Using CFD

- CFD analysis is reliable and has the capability to predict appropriately alteration in any course of the flow.
- CFD models have the ability to simulate complex flows in turbines or compressors.
- Utilizing CFD capabilities in the initial stage of compressor design for predicting deformity can save considerable time and reduce cost.
- The outcome from both steady and transient CFD simulations reveals that full scalability is nearly obtained by using commercial CFD codes.
- CFD is a widely used approach for designing of gas and steam turbines. The accurateness of CFD is rapidly growing with an increase in computational resources that catalyze high speed flow simulations.
- To control complex flow patterns in turbines and compressors, CFD has the competence to accurately calculate the nature of 3D flow.
- Modern turbomachinery designs are influenced entirely by CFD to model 3D blade sections. Modeling a single blade section is ample for the design and flow measurements of an entire stage using periodic boundary conditions.

2.2.1.2 Limitations of CFD in Turbomachinery

- 3D CFD analyses of turbomachinery blades are not yet sufficient enough to precisely predict the total losses observed in turbomachinery flows.
- Inflexible rules should essentially be laid out for CFD calculations so as to reduce the significance of mesh size in turbomachine models.
- Numerical errors are frequently observed in CFD calculations because of finite difference discretization schemes.

- Errors are encountered due to approximations in calculations where the true physics is still not established to predict complexities like turbulence flows.
- Unknown boundary conditions like tip clearance, leading/trailing edge, temperature profiles, and inlet/outlet pressure conditions still pose a major challenge.
- Several approximations and models are used in CFD and, as a result, errors originate during the discretization of the fundamental equations and also from truncation errors in turbulence models.
- The CFD model and the real geometry of the compressor and turbines have discrepancies as all geometrical features such as steps and fillets are not incorporated in the computational domain. Secondary air systems are snubbed in the case of a combined model of compressor/combustor/turbine while only their effects are considered using source terms (Denton and Dawes 1998).
- One of the most sensitive tasks in CFD is to accurately define boundary conditions that are crucial in turbomachinery applications.
- Accurate prediction of turbomachinery losses and understanding of its results are still challenging tasks that require great skill and expertise.
- Improved post-processing techniques are required for accurate interpretation of CFD results from unsteady flows.
- Accurate predictions of highly turbulent flows with affordable computational costs are still a major challenge to be addressed.

2.2.2 Basic Steps Involved in CFD Analysis

The essentials steps to be followed to develop a CFD model are explained in this section.

2.2.2.1 Problem Statement

The statement of turbomachinery problem normally defines the physics and identifies the data that is essential to express the problem mathematically and solve it using numerical approaches. A physical model is prepared from the information given in the problem statement.

2.2.2.2 Mathematical Model

The mathematical model is formulated from the physical behavior mentioned in the problem statement. Fluid flow equations are basically created from differential equations demonstrating the relationship linking flow variables and respective flow coordinates in space and time.

2.2.3 Governing Equations

The governing equations in fluid mechanics are the mass, momentum, and energy conservation equations. The Navier–Stokes equations are derived from the momentum equations. These equations are briefly explained in this section. The differential momentum equation is a nonlinear PDE in terms of dependent flow field variables like density, pressure, velocity, and so on. Therefore, the differential forms are most applicable for a complete analysis when field information is desired instead of integrated or averaged quantities. During the derivation of the differential equations of fluid motion, the numbers of equations are compared with number of unknown dependent fields when a solvable system of equations is established. The fundamental equations of fluid motion are derived based on the dynamical characteristics of the fluid, which is given by the following conservation laws:

1) Mass conservation law
2) Momentum-conservation law
3) Energy conservation law

2.2.3.1 Mass Conservation

Mathematical modeling of the physical principle for the conservation of mass leads to an integral or differential equation called an equation of continuity. The continuity equation shows that mass is conserved in a flow. The sum of all the fluid masses flowing inside and outside the control volume per unit time per unit volume must be equivalent to the rate of change of mass inside the control volume. The continuity equation for compressible flow is given by:

$$-\frac{\partial \rho}{\partial t} = \frac{\partial(\rho u)}{\partial x} + \frac{\partial(\rho v)}{\partial y} + \frac{\partial(\rho w)}{\partial z} \tag{2.3}$$

where u, v, and w are the velocity components in the x, y, and z directions, and ρ is fluid density.

The continuity equation applies to all fluids, compressible and incompressible flow, Newtonian or non-Newtonian fluids. It represents the law of conservation of mass at each point in a fluid and must therefore be satisfied at every point in a flow field. For incompressible flow, density is constant. The rate of change of density of fluid per unit time of a fluid particle will be zero. The continuity equation for incompressible flow can be simplified as:

$$\frac{\partial u}{\partial x} + \frac{\partial v}{\partial y} + \frac{\partial w}{\partial z} = 0 \tag{2.4}$$

2.2.3.2 Momentum Conservation

The momentum-conservation equation is developed from Newton's second law, the fundamental principle governing fluid momentum. However, the derivation is not stated in this book and can be found in various fluid dynamics books. The Navier–Stokes equations for a 3D, viscous, compressible flow are given next:

x-momentum equation:

$$\rho \underbrace{\left(\frac{\partial u}{\partial t} + u\frac{\partial u}{\partial x} + v\frac{\partial u}{\partial y} + w\frac{\partial u}{\partial z} \right)}_{\text{Inertial terms}}$$

$$= -\underbrace{\frac{\partial p}{\partial x}}_{\text{Pressure Gradient}} + \underbrace{\mu \left(\frac{\partial^2 u}{\partial x^2} + \frac{\partial^2 u}{\partial y^2} + \frac{\partial^2 u}{\partial z^2} \right)}_{\text{Viscous Terms}} + \underbrace{F_x}_{\text{Body Force Terms}} \tag{2.5}$$

y-momentum equation:

$$\rho \left(\frac{\partial v}{\partial t} + u\frac{\partial v}{\partial x} + v\frac{\partial v}{\partial y} + w\frac{\partial v}{\partial z} \right) = -\frac{\partial p}{\partial y} + \mu \left(\frac{\partial^2 v}{\partial x^2} + \frac{\partial^2 v}{\partial y^2} + \frac{\partial^2 v}{\partial z^2} \right) + F_y \tag{2.6}$$

z-momentum equation:

$$\rho\left(\frac{\partial w}{\partial t} + u\frac{\partial w}{\partial x} + v\frac{\partial w}{\partial y} + w\frac{\partial w}{\partial z}\right) = -\frac{\partial p}{\partial z} + \mu\left(\frac{\partial^2 w}{\partial x^2} + \frac{\partial^2 w}{\partial y^2} + \frac{\partial^2 w}{\partial z^2}\right) + F_z \quad (2.7)$$

where t is the temperature of the fluid, p is the fluid pressure, and μ is the dynamic viscosity of the fluid. The terms on the left-hand side are represented as advective terms and arrive from momentum fluctuations. These advective terms are opposed by the pressure gradient term $\left(\frac{\partial p}{\partial x}\right)$ followed by viscous forces that continuously force to slow down the flow, and F_x, F_y, and F_z are the body forces in x, y, and z coordinates.

The advective terms give the amount of the variation in velocity of one fluid component as it moves in space. The $\frac{\partial}{\partial t}$ term, known as the local derivative, gives the rate of change of fluid velocity in a fixed point in the flow field. The remaining three terms given in the inertial terms are clubbed together and are called the convective terms.

2.2.3.3 Energy Conservation

The fundamental principle that is applied in the derivation of the energy equation is the first law of thermodynamics and states that an increase of energy in a system is equal to heat added to that system plus the work done on it. However, only the final equation is given in this chapter. The conservation of energy equation has some similarities to that of the momentum equations and is given by:

$$\rho C_p\left(\frac{\partial T}{\partial t} + u\frac{\partial T}{\partial x} + v\frac{\partial T}{\partial y} + w\frac{\partial T}{\partial z}\right)$$

$$= \emptyset + \frac{\partial}{\partial x}\left[k\frac{\partial T}{\partial x}\right] + \frac{\partial}{\partial y}\left[k\frac{\partial T}{\partial y}\right] + \frac{\partial}{\partial z}\left[k\frac{\partial T}{\partial z}\right] + \left(u\frac{\partial p}{\partial x} + v\frac{\partial p}{\partial y} + w\frac{\partial w}{\partial z}\right) \quad (2.8)$$

where,

$$\emptyset = 2\mu\left[\left(\frac{\partial u}{\partial x}\right)^2 + \left(\frac{\partial v}{\partial y}\right)^2 + \left(\frac{\partial w}{\partial z}\right)^2 + 0.5\left(\frac{\partial u}{\partial y} + \frac{\partial v}{\partial x}\right)^2 + 0.5\left(\frac{\partial v}{\partial z} + \frac{\partial w}{\partial y}\right)^2\right.$$

$$\left. + 0.5\left(\frac{\partial w}{\partial x} + \frac{\partial w}{\partial y}\right)^2\right] - \frac{2}{3}\mu\left(\frac{\partial u}{\partial x} + \frac{\partial v}{\partial y} + \frac{\partial w}{\partial z}\right)^2 \quad (2.9)$$

where C_p is the specific heat at constant pressure and k denotes the thermal conductivity coefficient. The terms on the left side of Eq. (2.8) represent the rate of change of total energy per unit volume in the control volume and the rate of depletion of total energy by convection per unit volume through the control volume. The terms on the right side of the equation embody the heat generated per unit volume by external sources followed by the rate of heat lost by conduction per unit volume through the control volume. The next two terms signify the work done on the control volume per unit volume by the surface forces and body forces, respectively.

2.2.4 Turbulence Modeling

Turbulence is a leading topic in modern fluid dynamics research, and some of the best-known physicists have worked in this area during the last century. Among them are G. I. Taylor, Kolmogorov, Reynolds, Prandtl, von Karman, Heisenberg, Landau,

Millikan, and Onsagar. The first methodical work on turbulence was conducted by Osborne Reynolds in 1883. His research in flow through pipes demonstrated that the flow converts to a turbulent flow when the dimensionless ratio Reynolds number, $\mathrm{Re} = \rho VD/\mu$, exceeds a definite critical value. This dimensionless number, named after Osborne Reynolds by Sommerfeld, later proved to be the most important parameter that helps in deciding the flow behavior of viscous flows.

2.2.4.1 What is Turbulence?

As discussed earlier, when the Reynolds number rises, a three-dimensional vorticity fluctuations, dissipative, diffusive and random flow behavior can be observed. The velocity, as well as other flow properties, differs in a chaotic and random way. This flow regime is called turbulent flow. Turbulence is a natural phenomenon that occurs with high velocity gradients in fluids, causing perturbations in the flow domain as a function of time and space. Examples are aplenty in nature, such as smoke in the air, ocean waves, condensation of air on a wall, atmospheres of planets, and stormy weather. Turbulent flows are observed in flow through turbines, engines, compressors, combustion chamber, and so on.

2.2.4.2 Need for Turbulence Modeling

Turbulence modeling is the construction and use of a model to predict the effects of turbulence. Most flows found in engineering applications are turbulent, hence turbulent flow regimes are not just of theoretical importance. Fluid dynamic engineers necessarily require access to practical tools proficient in tackling various turbulence effects.

2.2.4.3 Reynolds-Averaged Navier–Stokes Equations

Turbulent flows can be formulated by averaging the Navier–Stokes system of equations over space and time. In 1895, Reynolds presented the first mathematical approach for prediction of turbulent flows. The methodology is established by decomposing the flow variables into two parts: a mean part and a fluctuating part. The conservation equations (Eqs. (2.5)–(2.7)) are then resolved for the mean values. The velocity components:

$$u_i = \overline{u}_i + \acute{u}_i \tag{2.10}$$

where $\overline{u}_i$ and $\acute{u}_i$ are the mean and fluctuating velocity components ($i = 1, 2, 3$). Likewise, for pressure and other scalar quantities:

$$\emptyset = \overline{\emptyset} + \acute{\emptyset} \tag{2.11}$$

where $\emptyset$ denotes a scalar such as pressure, energy, or species concentration. The mean values are achieved by an averaging technique. There are numerous ways of averaging flow variables: time averages, ensemble averages, spatial averages, and mass averages. However, only the time averaging of flow variables is explained here.

Let us assume variable f is the sum of the mean quantity $\overline{f}$ and fluctuating part $\acute{f}$, then f becomes,

$$f(\mathbf{x}, t) = \overline{f}(\mathbf{x}, t) + \acute{f}(\mathbf{x}, t) \tag{2.12}$$

where $\overline{f}$ is coined the time average of f,

$$\overline{f}(x, t) = \frac{1}{\Delta t} \int_t^{t+\Delta t} f(x, t)dt \tag{2.13}$$

with

$$\overline{\hat{f}} = \frac{1}{\Delta t} \int_t^{t+\Delta t} \hat{f}dt = 0 \tag{2.14}$$

In Eqs. (2.13) and (2.14), the time interval Δt is selected compatible with the time scale of the turbulent fluctuations for variable f as well as other variables within the physical domain. With time average for incompressible flows, the conservation equations can be written as follows:

$$\frac{\partial \overline{u}_i}{\partial x_i} = 0 \tag{2.15}$$

In Eq. (2.15), tensor notation u_i denotes a velocity component ($u_i = [u_1 + u_2 + u_3]^T$) and x_i denotes a coordinate direction.

The momentum equation for x-momentum can be expressed as:

$$\rho \left(\frac{\partial \overline{u}_i}{\partial t} + \overline{u}_j \frac{\partial \overline{u}_i}{\partial x_j} \right) = -\frac{\partial \overline{p}}{\partial x_i} + \frac{\partial}{\partial x_j}(\overline{\tau}_{ij} - \rho \overline{\hat{u}_i \hat{u}_j}) \tag{2.16}$$

This system of equations is called the Reynolds-averaged Navier–Stokes (RANS) equations. These equations of turbulence are similar to the Navier–Stokes equations (Eqs. (2.5)–(2.7)) without an additional term

$$\tau_{ij}^R = -\rho \overline{\hat{u}_i \hat{u}_j} = -\rho \left(\overline{u_i u_j} - \overline{u}_i \overline{u}_j \right) \tag{2.17}$$

which constitute the Reynolds Stress Tensor, τ_{ij}, and symbolizes the transfer of momentum due to turbulent fluctuations.

The laminar viscous stresses are expressed as,

$$\overline{\tau}_{ij} = 2\mu \overline{S}_{ij} = \mu \left(\frac{\partial \overline{u}_i}{\partial x_j} + \frac{\partial \overline{u}_j}{\partial x_j} \right) \tag{2.18}$$

2.2.4.4 Turbulence Closure Models

In order to analyze turbulent flows using the RANS equations, it is necessary to use turbulence closure models to predict the Reynolds stresses and scalar transport terms. The first-order closures signify the simplest method to approximate the Reynolds stress terms in the RANS equations. For a turbulence model to be used in a general CFD code, the model must have wide applicability, reasonable accuracy, and simplicity. From the large selection of first-order closure models, the commonly used models in the current state-of-the-art simulations are explained next. All three models can be employed on both structured and unstructured meshes. First of all, the one-equation model of Spalart and Allmaras (1992) is discussed. Second, the renowned k-ε two-equation model is presented. Finally, the k-ω SST (Shear Stress Transport) two-equation model is explained.

2.2.4.4.1 One-Equation Model: Spalart–Allmaras Model

The Spalart–Allmaras (SA) model is an example of a one-equation turbulence model (Spalart and Allmaras 1992). This model addresses the flow equation of an eddy viscosity variable. Practically accurate predictions of turbulent flows that have adverse pressure gradients can be obtained. Additionally, smooth transition from laminar to turbulent flow at user specified locations are possible. In the SA model, the equation at one location does not depend on the results obtained at another location. Hence, this model can be easily employed on structured or unstructured grids. It converges quickly to steady state and needs only adequate grid concentration at the near wall section. Robustness is another advantage of this model. The one-equation SA model was initially intended for aerospace models containing wall-bounded flows that gave good results for boundary layers exposed to adverse pressure gradients. It is also attaining acceptance in turbomachinery applications.

The transport equation for eddy viscosity variable $\tilde{v}$ is given by:

$$\frac{\partial}{\partial t}(\rho\tilde{v}) + \frac{\partial}{\partial x_i}(\rho\tilde{v}u_i) = G_v + \frac{1}{\sigma_{\tilde{v}}}\left[\frac{\partial}{\partial x_j}\left\{(\mu + \rho\tilde{v})\frac{\partial\tilde{v}}{\partial x_j}\right\} + C_{b2}\rho\left(\frac{\partial\tilde{v}}{\partial x_j}\right)^2\right] - Y_v + S_{\tilde{v}}$$

(2.19)

where G_v is the turbulent viscosity production term, Y_v is the turbulent viscosity destruction term, $\sigma_{\tilde{v}}$ and C_{b2} are constants, and μ is the absolute viscosity.

The first term on the left-hand side of the SA equation is the rate of change of $\tilde{v}$. The second term is the transport of $\tilde{v}$ by convection. On the right-hand side, the first term is the rate of production of $\tilde{v}$. The next term is the transport of $\tilde{v}$ by turbulent diffusion. The fifth term is the rate of dissipation of $\tilde{v}$. $S_{\tilde{v}}$ is a user-defined source term.

The turbulent viscosity (μ_t) can be calculated by:

$$\mu_t = \rho\tilde{v}\frac{X^3}{X^3 + C_{v1}^3}$$

(2.20)

where $X \equiv \frac{\tilde{v}}{v}$, and the production term is given as,

$$G_v = C_{b1}\rho\tilde{S}\tilde{v}$$

(2.21)

where $\tilde{S} \equiv S + \frac{\tilde{v}}{k^2 d^2}f_{v2}, f_{v2} = 1 - \frac{X}{1+Xf_{v1}}$ and $S \equiv \sqrt{2\Omega_{ij}\Omega_{ij}}$, and C_{b1} and k are constants, d is the distance from the wall, S is a scalar measure of the deformation tensor, and Ω_{ij} is the mean rate-of-rotation tensor, which is defined as:

$$\Omega_{ij} = \frac{1}{2}\left(\frac{\partial u_i}{\partial x_j} - \frac{\partial u_j}{\partial x_i}\right)$$

(2.22)

The turbulent destruction term can be written as:

$$Y_v = C_{w1}\rho f_w\left(\frac{\tilde{v}}{d}\right)^2$$

(2.23)

where $f_w = g\left[\frac{1+C_{w3}^6}{g^6+C_{w3}^6}\right]^{1/6}; g = r + C_{w2}(r^6 - r); r \equiv \frac{\tilde{v}}{\tilde{S}\kappa^2 d^2}$

The model constants are: $C_{b1} = 0.1355$, $C_{b2} = 0.622$, $\sigma_{\tilde{v}} = 2/3$, $C_{v1} = 7.1$, $C_{w2} = 0.3$, $C_{w3} = 2.0$, and $\kappa = 0.4187$. And, C_{w1} is given as:

$$C_{w1} = \frac{C_{b1}}{\kappa^2} + \frac{(1 + C_{b2})}{\sigma_{\tilde{v}}} \tag{2.24}$$

2.2.4.4.2 Two-Equation Model

Two-equation turbulence models determine both a turbulent length and time scale by solving two transport equations.

The k-ε Model The *k-epsilon* or *k-ε* turbulence model is one of the widely used two-equation turbulence models in industry and academia. It is built on the resolution of two equations; that is, the turbulent kinetic energy equation and the turbulent dissipation rate equation. It was proposed by Launder and Spalding (1974) and has become the pillar of engineering turbulent flow calculations ever since. The standard $k-\varepsilon$ model has two model equations, one for k (turbulence kinetic energy) and one for ε (turbulence dissipation rate) and is given next as follows:

k equation:

$$\frac{\partial}{\partial t}(\rho k) + \frac{\partial}{\partial x_i}(\rho k u_i) = \frac{\partial}{\partial x_j}\left[\left(\mu + \frac{\mu_t}{\sigma_k}\right)\frac{\partial k}{\partial x_j}\right] + G_k + G_b - \rho\varepsilon - Y_M + S_k \tag{2.25}$$

ε equation:

$$\frac{\partial}{\partial t}(\rho\varepsilon) + \frac{\partial}{\partial x_i}(\rho\varepsilon u_i) = \frac{\partial}{\partial x_j}\left[\left(\mu + \frac{\mu_t}{\sigma_\varepsilon}\right)\frac{\partial\varepsilon}{\partial x_j}\right] + C_{1\varepsilon}\frac{\varepsilon}{k}(G_k + C_{3\varepsilon}G_b) - C_{2\varepsilon}\rho\frac{\varepsilon^2}{k} + S_\varepsilon \tag{2.26}$$

In words, these equations can be expressed as:

Rate of change of k or ε + Transport of k or ε by convection

= Transport of k or ε by diffusion + Rate of production of k or ε

− Rate of destruction of k or ε + user − defined source terms (2.27)

Here, G_k is the generation rate of turbulent kinetic energy as a result of the mean velocity gradients, G_b is the generation rate of turbulence kinetic energy thanks to buoyancy effects, Y_M signifies the influence of the fluctuating dilatation in compressible turbulence to the overall dissipation rate, $C_{1\varepsilon}$, $C_{2\varepsilon}$, and $C_{3\varepsilon}$ are model constants, S_k and S_ε are user-defined source terms, and σ_k and σ_ε are turbulent Prandtl numbers for k and ε, respectively.

The turbulent viscosity term (μ_t) is calculated by:

$$\mu_t = \rho C_\mu \frac{k^2}{\varepsilon} \tag{2.28}$$

where the model constants have the values: $C_\mu = 0.09$, $C_{1\varepsilon} = 1.44$, $C_{2\varepsilon} = 1.92$, $\sigma_k = 1.0$, and $\sigma_\varepsilon = 1.3$.

The k-ω Model The kinematic eddy viscosity μ_t in the $k-\varepsilon$ model, is stated as the product of velocity scale $\mu = \sqrt{k}$ and the length scale $l = k^{3/2}/\varepsilon$. The rate of turbulence kinetic energy dissipation (ε) is not the only potential length scale defining a variable. As a matter of fact, there are many other two-equation models that have been proposed. The best remarkable alternative is the k–ω model proposed by Wilcox (1988), which uses the turbulence frequency omega $\omega = \varepsilon/k$ (dimensions $s-1$) as the second variable. The length scale becomes $l = \sqrt{k}/\omega$. The turbulence kinetic energy and the specific dissipation rate is obtained from the following transport equations:

k equation:

$$\frac{\partial}{\partial t}(\rho k) + \frac{\partial}{\partial x_i}(\rho k u_i) = \frac{\partial}{\partial x_j}\left[\Gamma_k \frac{\partial k}{\partial x_j}\right] + G_k - Y_k + S_k \tag{2.29}$$

ω equation:

$$\frac{\partial}{\partial t}(\rho \omega) + \frac{\partial}{\partial x_i}(\rho \omega u_i) = \frac{\partial}{\partial x_j}\left[\Gamma_\omega \frac{\partial \omega}{\partial x_j}\right] + G_\omega - Y_\omega + S_\omega \tag{2.30}$$

In words, the equations represent:

Rate of change of k or ω + Transport of k or ω by convection

= Transport of k or ω by turbulent diffusion + Rate of production of k or ω

− Rate of dissipation of k or ω + user − defined source terms $\qquad$ (2.31)

Here, G_ω is the generation rate of ω, Y_k and Y_ω denote the dissipation of k and ω due to turbulence, Γ_k and Γ_ω denote the effective diffusivity of k and ω, respectively, and S_k and S_ω are user-defined source terms.

The effective diffusivities are defined by:

$$\Gamma_k = \mu + \frac{\mu_t}{\sigma_k} \ and \ \Gamma_\omega = \mu + \frac{\mu_t}{\sigma_\omega} \tag{2.32}$$

where σ_k and σ_ω are turbulent Prandtl numbers for k and ω, respectively, and, the turbulent viscosity μ_t can be calculated by combining k and ω as $\mu_t = \alpha^* \frac{\rho k}{\omega}$, where α^* is the damping coefficient of turbulent viscosity triggering a low Re modification.

Shear Stress Transport (SST) Model Menter (1994) established the SST model to effectively merge the free-stream objectivity of the k–ε model in the far field with the robust and accurate prediction of the k–ω model in the near wall region. The k equation is the same as in Wilcox's original k–ω model, but the ε equation is transformed into a ω equation by substituting $\varepsilon = k\omega$. The SST model incorporates a damped cross-diffusion derivative term in the ω equation. Thus, the transport equations for the k–ω SST model can be written as:

k equation:

$$\frac{\partial}{\partial t}(\rho k) + \frac{\partial}{\partial x_i}(\rho k u_i) = \frac{\partial}{\partial x_j}\left[\Gamma_k \frac{\partial k}{\partial x_j}\right] + \widetilde{G_k} - Y_k + S_k \tag{2.33}$$

ω equation:

$$\frac{\partial}{\partial t}(\rho \omega) + \frac{\partial}{\partial x_i}(\rho \omega u_i) = \frac{\partial}{\partial x_j}\left[\Gamma_\omega \frac{\partial \omega}{\partial x_j}\right] + G_\omega - Y_\omega + D_w + S_\omega \tag{2.34}$$

where $\widetilde{G_k}$ is the generation rate of turbulence kinetic energy due to mean velocity gradients, calculated from G_k, and D_w represents the cross-diffusion term.

The turbulent viscosity μ_t is defined in terms of k and ω as follows:

$$\mu_t = \frac{\rho k}{\omega} \frac{1}{max\left[\frac{1}{\alpha^*}, \frac{SF_2}{a_1\omega}\right]}$$

(2.35)

2.2.4.5 Large Eddy Simulation (LES)

The LES is an alternate approach toward realizing more capable turbulent flow calculations. A far more refined mesh than the RANS system of equations is used in LES. The precision of LES in terms of performance and ability is someplace between RANS and the Direct Numerical Simulation (DNS), which is discussed in Subsection 2.2.4.6.

Turbulent flows are portrayed by eddies in different length and time scales. The largest eddies are normally equivalent in size to the characteristic length of the average flow. The lowest scales are accountable for the liberation of turbulence kinetic energy. There are two major steps involved in LES analysis: filtering and subgrid scale (SGS) modeling. Filtering is used for solving the large eddies involved while the SGS model addresses the small eddies that are unresolved in the filtering method.

Large unsteady turbulent eddies are proposed using this technique, though small scale dissipative turbulent eddies are modeled. The basic theory is that the smaller turbulent eddies are not dependent on coordinate system orientation and will always act in a statistically parallel and expectable manner, regardless of the turbulent flow field.

The main limitation of LES lies in the wall boundary layers that require high-resolution grids. Even big eddies convert to small eddies near the wall and need a Reynolds number dependent resolution.

2.2.4.6   Direct Numerical Simulation (DNS)

In DNS, the Navier–Stokes equations are solved directly with refined meshes capable of resolving all turbulent length scales. These simulations calculate the mean flow along with all turbulent velocity fluctuations. DNS has potential for simulation of complex turbulent flows, and algorithms for solving 3D continuity and Navier–Stokes equations are well established. These calculations come with high costs in terms of computational resources, so this approach is not commonly applied for industrial flow computations.

DNS is expected to play a progressively important part in turbulence research in the immediate future. DNS methods need extremely fine, fully 3D meshes, huge computers, and a massive amount of CPU time. At present computational capabilities, DNS results are not yet realizable for highly turbulent flows in engineering applications.

2.2.5   Boundary Conditions

No numerical scheme is complete without proper ways of imposing boundary conditions. Suitable boundary conditions are mandatory to achieve accuracy CFD solution. Boundary conditions are stated on every edge in the 2D CFD domain and on every face in the 3D domain. Incorrect specifications of boundary conditions result in solution instability, non-convergence of solutions, and/or convergence to inaccurate results. Numerous types of boundary conditions exist and the most significant ones are discussed next.

2.2.5.1 Inlet/Outlet Boundary Conditions

Fluid enters the computational domain through an inlet and leaves the domain through an outlet specified on the geometry. These boundary conditions are normally classified as either velocity or pressure specified conditions. The velocity of the incoming flow is specified along the inlet edge/face at a velocity inlet. The temperature properties of the inward flow should be specified if energy equations are being solved. Same goes in case of turbulence, if turbulence is involved then properties of turbulence on the inward flow also should be defined. If a pressure inlet is specified, the total pressure is stated across the inlet edge/face (e.g., flow entering the domain from far field and the ambient pressure is well known or from a pressurized tank of known pressure). Pressure outlet is defined where the fluid flows out of the domain. Flow properties like temperature and turbulent intensities are also stated at pressure inlets and outlets.

2.2.5.2 Wall Boundary Conditions

The wall boundary condition is the simplest among boundary conditions. Since fluids cannot traverse through a solid wall, the velocity component normal to the wall is set to zero along an edge or face where the wall boundary conditions are set. Additionally, due to the no-slip condition of the wall, the velocity tangential component at a fixed wall is set to zero as well. Wall heat flux or wall temperatures are specified if the energy equation is invoked. Moving walls and walls with definite shear stresses are also possible for modeling in a majority of the CFD models available.

2.2.5.3 Periodic/Cyclic Boundary Conditions

Periodic flow arises where the physical geometry and the expected flow pattern have a periodically repeating nature. Such flows are witnessed in multiple applications like flows inside heat exchanger channels and flow between the blades of a turbomachine. It occurs when the geometry includes repetition. Flow variables defined on one face of a periodic boundary are mathematically connected to the second face of an identical boundary. Therefore, flow exiting one periodic boundary can be thought of as entering another periodic boundary with identical properties (temperature, velocity, pressure, etc.). The computational domain is reduced when the periodic boundary condition is applied, thus conserving computer resources (Figure 2.3).

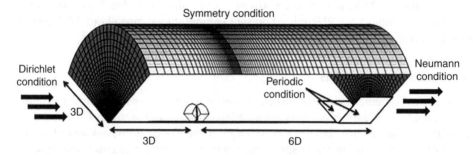

Figure 2.3 Mesh and periodic boundary conditions in a 3D axial flow problem. Source: Reprinted from Lee et al. (2012), (figure 2 from original source) with permission from Elsevier.

Figure 2.4 Symmetry boundary condition.

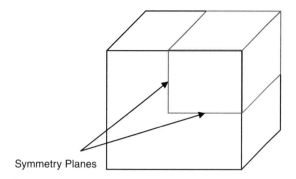

Symmetry Planes

2.2.5.4 Symmetry Boundary Conditions

Symmetry boundary conditions are applied when the physical geometry and the expected flow pattern have mirror symmetry. They are used to decrease the size of the computational domain to a symmetric section of the complete system. The symmetric boundary condition is shown in Figure 2.4.

2.2.6 Moving Reference Frame (MRF)

An MRF is a relatively simple, robust, and efficient steady-state CFD modeling technique used to simulate rotating machinery. An MRF assumes that an assigned volume has a constant speed of rotation and the non-wall boundaries are surfaces of revolution. Strictly speaking, the complete rotor domain is expected to be rotating at the angular velocity as that of the rotor. As all turbomachinery flows are fundamentally unsteady in nature, the MRF model simplifies these difficulties in CFD. It is simpler to define appropriate boundary condition calling for less time and computational resources and, consequently, this is the desired choice in engineering practices over other models. The MRF model has a variety of application in systems like turbomachinery, electric motors, generators, mixing equipment, and rotating passages as well as air and land vehicle motions.

When a MRF is initiated, additional acceleration terms like Coriolis and centripetal acceleration (which arise because of the alteration from stationary to the MRF) are

Figure 2.5 Stationary and moving reference frames. Source: Reprinted from Shi et al. (2015), (figure 4 from original source) with permission from Elsevier.

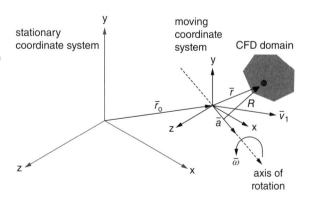

added to the equations of motion (Figure 2.5). The flow around moving parts can be predicted by resolving these equations in the steady state.

2.2.7 Verification and Validation

The verification calculation governs whether the programming and computational operation of the theoretical model is correct or not. It inspects the mathematics in the models by comparing the CFD solution to exact analytical results. It also examines errors arising from computer programming.

The verification calculation examines whether the computational models have correctly implemented the conceptual models and also whether the resulting code can be accurately used for further studies. This is done by identifying and quantifying the errors in the model development and the solution. The two phases of verification are calculation verification and CFD code verification. The intentions of verifying a code is to find and eradicate errors in the code. The intention of verifying a calculation is to determine the accurateness of a calculation.

Validation evaluation governs whether the computational model accords with the physical reality or not. It inspects the science in the models by comparing it to experimental data. This is done by identifying and quantifying error and uncertainty with an assessment of simulation and experimental results. The experiment results also might contain bias and random errors that must be properly computed and acknowledged as part of the dataset. The accurateness necessary for validation is reliant on the application and, hence, the validation needs to be accommodating to allow different levels of accuracy.

2.2.8 Commercial CFD Software

CFD analysis tools are now commercially developed and are becoming quite predominant in most industries. Sometimes an entire suite, including pre- and post-processors, is provided by a single software developer. Several of these commercial CFD software packages are specially designed to address for turbomachinery applications and some of them have the competency to not only treat blade rows but solve for other associated configurations including jets and nozzles.

For several years, ANSYS Fluent®, ANSYS CFX®, and Star-CCM+® have been dominating the CFD industry with their state-of-the-art tools. ANSYS Fluent appears to dominate the electronic and industrial product markets while Star-CCM+ is widely accepted in the aerospace, automotive, and energy industries. ANSYS Fluent also offers explicit turbomachinery post-processing features that aid in accurate result analysis.

ANSYS CFX is another revved-up CFD software tool that provides accurate and reliable results, robustly and quickly across a wide range of CFD and turbomachinery applications. ANSYS CFX is renowned for its exceptional robustness, accuracy, and speed with rotating equipment such as compressors, fans, pumps, and gas turbines. Turbo mode in CFX-Pre® is an expert mode enabling the user to conduct turbomachinery simulations, such as compressors or turbines with ease. Each section of a piece of rotating equipment can be simply constructed by choosing some basic parameters and a suitable mesh file, then the different boundary conditions associated with the section and interfaces between the components are generated automatically (ANSYS CFX Solver Modelling Guide 2013).

Another popular turbomachinery CFD software is the FINE™/Design3D® code developed by NUMECA. FINE/Design3D is an integrated environment for the design and optimization of turbomachinery channels and blades. An optimization system requires a parametric modeler to define a large design space, an automatic and fast grid generator, an accurate and robust CFD solver, and an efficient optimization kernel (ANSYS Fluent 12.0 2009).

The broad developments observed in turbomachinery design in recent years depict a continuing impulsion for better component performance and also the ongoing excellence in CFD modeling competencies. The prime factor leading the application of CFD modeling to turbomachinery component design is the computational capability of the workstation on which the CFD software is installed. As parallel processing technologies improve and workstation processor speeds up, more advanced CFD analysis tools can be accommodated in the turbo-components design process (Pinto et al. 2016). Nevertheless, decision making must be carried out on the basis of several other factors than merely on the capability of the CFD package to analyze complex turbomachinery related flows. The cost associated with purchasing or leasing commercial software must be analyzed before moving on to the CFD approach.

2.2.9 Open Source Codes

Open source CFD codes are also known as OFF (Open source Finite volume Fluid dynamics code). Their aim is to numerically solve steady and unsteady compressible Navier–Stokes equations of fluid transport by using FVM practices (Zaghi 2014). There are so many OFF codes available due to increased usage of CFD. These codes are available under the GNU General Public License, so anyone can use them and modify them without any restrictions, thus promising the distribution of scientific knowledge between researchers. Moreover, free software often is of greater quality compared to its commercial counterparts, especially for scientific applications. The limitations of open source codes are that they need coding skills and the complete understanding of flow physics, among other things.

A list of the most widely used open source CFD codes is given in Table 2.1.

2.2.9.1 OpenFOAM

OpenFOAM can simulate a vast number of problems in CFD because of its wide range of solvers. OpenFOAM consists of meshing tools like blockMesh and snappy HexMesh. BlockMesh supports meshing of simple geometries while the snappy HexMesh can support complex geometries (Jasak and Beaudoin 2011). There are options to import mesh files from GAMBIT, Star-CD and ANSYS. There are various mesh conversion tools to convert the mesh files from other formats to those that can be read by OpenFOAM.

The simulation of turbomachines is significantly different from ordinary simulations like internal flow through pipes, external aerodynamics CFD, heat transfer problems, or wall-bounded turbulent flows. The unsteady nature of flow along with the complex wake to blade interaction makes the CFD of turbomachinery challenging. To make Open-FOAM suitable for simulating turbomachines, the rotation and the interface between the stationary and rotating parts are the difficulties that one has to address (Page and Beaudoin 2007).

Table 2.1 Open source CFD codes.

Code name	Explanation
OpenFOAM	This is the most extensively used open source software. which is versatile and can be applied to various engineering and science applications. It is an object-oriented programming developed in C++ library for CFD numerical modeling, with second order accurate FVM discretization, second order discretization in time, provision for parallel computing and effective linear system solvers. It includes over 80 solver applications that can simulate specific problems in engineering.
REEF3D	REEF3D strongly concentrates on hydraulic, offshore, coastal, and marine CFD. Complicated free surface flows are computed easily with the use of level-set method. The model is employed in highly modular C++ code and the source code is accessible freely under the GPL license. REEF3D is the best tool for high-performance CFD modeling thanks to the MPI library which used for parallelization. The model is presently under development and further new features are affixed rapidly. Their goal is to transform from a fully transitional research tool to powerful engineering software.
Code Saturne	Code Saturne was developed by EDF (Électricité de France) with the purpose of research and industrial applications. It solves 2D and 3D Navier–Stokes equations, 2D axisymmetric flows, laminar or turbulent, incompressible or compressible, isothermal or adiabatic, steady or unsteady with scalar transport if needed. It can also support sliding meshes. Several turbulence models are available, from RANS to LES models. The structure of the code remains the same over the versions with the major change from 1.0× to 2.0× being the upgrade from FORTRAN 77 to FORTRAN 90.
Gerris	Gerris was created by Stéphane Popinet and is supported by NIWA (National Institute of Water and Atmospheric research) and Institute Jean le Rond D'Alembert. It solves the time-dependent incompressible variable-density Navier–Stokes equations, Volume of Fluid advection scheme for interfacial flow, linear, and nonlinear shallow-water equations with second-order accuracy in space and time. It also supports Adaptive mesh refinement. But it is not used for compressible flows.
SU2	SU2 is created in C++ for solving PDE and PDE-constrained optimization simulations. The SU2 tool was initially built in consideration with CFD and aerodynamic shape optimization, but recently it has been developed to solve new circles of governing equations like chemically reacting flows, electrodynamics, and few others. SU2 is under constant expansion at the Aerospace Design Lab under at Stanford University.

References

Anderson, J.D. (1995). *Computational Fluid Dynamics: The Basics with Applications*. New York: McGraw-Hill.

ANSYS CFX (2013). *Solver Modelling Guide*. Canonsburg: ANSYS, Inc.

ANSYS Fluent 12.0 (2009). *Theory Guide*. Lebanon, NH, USA: ANSYS Inc.

Çengel, Y.A. and Cimbala, J.M. (2006). *Fluid Mechanics: Fundamentals and Applications*. New York, NY: McGraw-Hill Education.

Chung, T.J. (2015). *Computational Fluid Dynamics*. Cambridge: Cambridge University Press.

Denton, J.D. and Dawes, W.N. (1998). Computational fluid dynamics for turbomachinery design. *Proceedings of the Institution of Mechanical Engineers, Part C: Journal of Mechanical Engineering Science* 213 (2): 107–124.

Jasak, H., and Beaudoin, M. (2011). OpenFOAM Turbo Tools: From General Purpose CFD to Turbomachinery Simulations. ASME-JSME-KSME 2011 Joint Fluids Engineering Conference: Volume 1, Symposia – Parts A, B, C, and D.

Launder, B.E. and Spalding, D.B. (1974). The numerical computation of turbulent flows. *Computer Methods in Applied Mechanics and Engineering* 3 (2): 269–289.

Lee, J.H., Park, S., Kim, D.H. et al. (2012). Computational methods for performance analysis of horizontal axis tidal stream turbines. *Applied Energy* 98: 512–523. https://doi.org/10.1016/j.apenergy.2012.04.018.

Menter, F.R. (1994). Two-equation eddy viscosity turbulence models for engineering applications. *AIAA Journal* 32: 1598–1605.

Page, M., and Beaudoin, M. (2007). Adapting OpenFOAM for turbomachinery applications. Second OpenFOAM Workshop in Zagreb, Croatia.

Pinto, R.N., Afzal, A., D'Souza, L.V. et al. (2016). Computational fluid dynamics in turbomachinery: a review of state of the art. *Archives of Computational Methods in Engineering* 24 (3): 467–479.

Shi, S., Zhang, M., Fan, X., and Chen, D. (2015). Experimental and computational analysis of the impeller angle in a flotation cell by PIV and CFD. *International Journal of Mineral Processing* 142: 2–9. https://doi.org/10.1016/j.minpro.2015.04.029.

Spalart, P. R., and Allmaras, S. R. (1992). A One-Equation Turbulence Model for Aerodynamic Flows, AIAA Paper 92–0439.

White, F.M. (2017). *Fluid Mechanics*. New Delhi, India: McGraw-Hill Education.

Wilcox, D.C. (1988). Reassessment of the scale-determining equation for advanced turbulence models. *AIAA Journal* 26 (11): 1299–1310.

Zaghi, S. (2014). OFF, open source finite volume fluid dynamics code: a free, high-order solver based on parallel, modular, object-oriented Fortran API. *Computer Physics Communications* 185 (7): 2151–2194.

3

Optimization Methodology

3.1 Introduction

The meaning of "optimization" in the Cambridge Dictionary (Vale et al. 1996) is "the act of making something as good as possible." In our day-to-day life, we perform optimizations without knowing any complex mathematics behind them. We find the shortest and easiest route to reach a shop to purchase milk or optimize expenditures before buying a car or an apartment. Engineering systems may be simple or complex and then optimized to find the best possible ways to construct a bridge, an aircraft, or a system of a chemical process. Systems can have hundreds of parameters or variables that can be modified to achieve goals or objectives. As the number of variables and/or objectives increases, the problem turns into a more complex problem. An experienced engineer can somehow guess an optimal design, but the "guess" can fail as the system complexity varies. Over the last few decades, computer performance has been increased several-fold and complex designs are now initially computed by numerical techniques such as computational fluid dynamics (CFD), finite element methods (FEM) or other techniques. Thus, several mathematical optimization techniques come into play to integrate optimization with numerical techniques. The optimization is gradually becoming a central theme of modern engineering practices. Several parameters or variables can be varied or put under constraints so that a certain goal can be achieved. The goals or objectives can be maximized, minimized, or constrained. Goals can be singles or multiples. If a problem has a single goal, it is called a single-objective problem, while the other is called a multi-objective problem.

Engineering system design and optimization through simulations can take a fraction of a second to several days. Hence, there are mathematical approaches to connect a design formula to an optimization algorithm. In surrogate-based optimization methods, a low-fidelity model or surrogate model construction is made to reduce total computational time.

There are also several direct gradient-based optimization methods such as finite difference sensitivity analysis, automatic differentiation, and adjoint variable methods. Local and global sensitivity analysis gives information of the effects of variables on the objectives. In the automatic differentiation method, the problem is transformed into a function that performs the same way as in the original problem with the same accuracy and efficiency. The adjoint method for a constraint optimization algorithm gives faster convergence, and is introduced in Section 3.10.

Design Optimization of Fluid Machinery: Applying Computational Fluid Dynamics and Numerical Optimization, First Edition. Kwang-Yong Kim, Abdus Samad, and Ernesto Benini.
© 2019 John Wiley & Sons Singapore Pte. Ltd. Published 2019 by John Wiley & Sons Singapore Pte. Ltd.

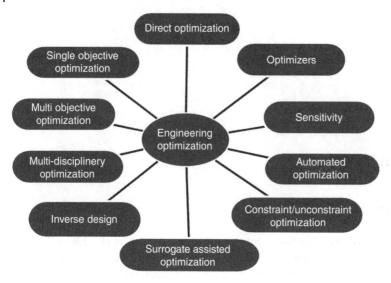

Figure 3.1 The terms commonly used in engineering optimization.

In this chapter, different optimization techniques are discussed mostly for surrogate-based optimization. The target of this book relates to fluid machinery optimization. The techniques used in chemical or management processes are avoided.

Figure 3.1 shows the terms commonly used in engineering optimization.

3.1.1 Engineering Optimization Definition

A decision-making process in engineering depends on a single parameter or multiple parameters that affect the system performance. The optimization methodology used in the process deals with choosing an alternative among a set of designs. Generally, an optimization process can be expressed as

$$\text{Minimize } f(x) \tag{3.1}$$

Subject to decision variable set, $x \in \Omega$ (Ω is a subset of real space R^n).

An optimization procedure based on surrogate modeling is represented in Figure 3.2. Primarily, a design space is defined and several samples are nominated by Design of Experiments (DOEs) (Myers et al. 2016). Next, the designs are evaluated through the experimental or numerical approach and the optimization algorithms are implemented. In many other cases, the design space may not be defined, rather, the system is optimized considering only one parameter at a time. This approach is time consuming.

3.1.2 Design Space

When some variables are selected for modifications in a system to produce an optimal shape, the computations are to be done in a limited range of design space. For example, a symmetric airfoil profile can have an NACA0010 to NACA0020 profile. If a design exceeds the limits, the design may not be structurally strong or will be unfeasible. Hence, a design space is defined by the lower limit and upper limit of the variables selected. The design space can be expressed mathematically as:

Figure 3.2 A surrogate-based optimization procedure. Source: With kind permission from Springer Science+Business Media: Samad and Kim 2008 (figure 3 from original source), © 2008.

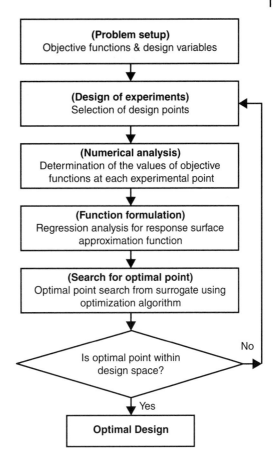

$$x_i^l \leq x_i \leq x_i^u \tag{3.2}$$

Searching a proper design space is a critical issue in engineering design. The lower or upper limit of a variable may be selected through designers' prior experiences or through some random calculations or through design checks. Any variable with the lowest value should produce a feasible design, and it should achieve some design objectives.

Now, take an example of two variables (V1 and V2) that are in a feasible range. The combination of the ranges may not produce a feasible design space. The lower limit of V1 and the upper limit of V2 may create a completely absurd design space.

When a designer is selecting a design space, it is necessary to check individual variable limits as well as combined limits. It is recommended to initially check the combined limits of the variables as to whether they work well or not. Subsequently, the designer can select design points from the design space.

3.1.3   Design Variables and Objectives

A turbomachinery system may have hundreds of design parameters that can alter objective function values. It is the responsibility of the designer to select the most significant variables. An increasing number of design variables increases the number

Figure 3.3 Axial compressor.

of design evaluations. To decrease the number of variables, it is necessary to critically evaluate the design variables. Design analysis should be based on feasibility, flow physics, constraints, and so on. If altering the value of design variables does not affect the objective function, then those variables can be ignored and the variables having larger effects can be selected. Conducting sensitivity analysis of variables will give some clue to variable selection.

For example, let us consider an example of a turbomachinery blade (Figure 3.3). An axial compressor can have "*n*" number of blades and the blade shape can affect the compressor performance. There are parameters that can be changed and the blade performance can be altered. The specification of the compressor is given in Table 3.1. Details of the compressor and its optimization will be discussed in another chapter.

Here, only two parameters are taken (blade lean and blade sweep). The blade shape can be defined in terms of blade lean and blade sweep. The lean modifies the blade

Table 3.1 Specification of an axial compressor.

Mass flow rate, kg s^{-1}	20.19
Rotational speed, rpm	17,190
Number of rotor blades	36
Hub-tip ratio at inlet	0.7
Inlet tip relative Mach no.	1.4
Inlet hub relative Mach no.	1.13
Tip solidity	1.29
Pressure ratio	2.106
Aspect ratio of rotor	1.19

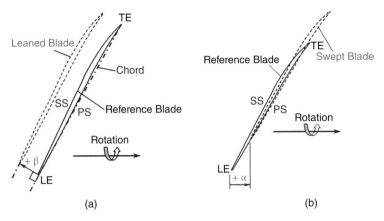

Figure 3.4 Classification of design variables (LE: Leading edge, TE: trailing edge, SS: Suction surface, PS: pressure surface). (a) Blade lean and (b) blade sweep. Source: Reproduced with permission from Samad et al. (2008a), (figures 3, 4 from original source) Copyright © 2008 by the American Institute of Aeronautics and Astronautics, Inc.

and moves the tip profile perpendicular to the chord line and the blade bends toward pressure or the suction side; on the other hand, the sweep modifies the blade by changing the blade tip along the chord line (Figure 3.4). Hence, a designer should modify only one or both of the parameters or variables to a get higher efficiency of pressure ration. The design space is shown in Table 3.2. The variables are selected using a sampling technique.

The samples are given in Table 3.3. The last two columns are kept blank and are filled with CFD calculation results. Once the CFD results are obtained, the table is used to construct surrogates or approximating functions. A search algorithm finds the optimal point from the surrogates. The results are presented after CFD simulations (Figure 3.5). The reference (*ref*) and the improved-optimized results are shown.

The objectives are defined as:

Efficiency

$$\eta_{ad} = \frac{(P_{total,exit}/P_{total,inlet})^{(k-1)/k} - 1}{T_{total,exit}/T_{total,inlet} - 1} \tag{3.3}$$

Pressure ratio

$$F_P = P_{total,exit}/P_{total,inlet} \tag{3.4}$$

Table 3.2 Design space.

Variables	Lower limit	Upper limit
Sweep, α	0.0	0.25
Lean, β	−0.036	0.000

Table 3.3 Design table or samples.

S.No.	V_1 limit	V_2 limit	Efficiency	Pressure ratio
1	0.0	−0.036		
2	0.0	−0.018		
3	0.0	0.000		
4	0.13	−0.036		
5	0.13	−0.018		
6	0.13	0.000		
7	0.25	−0.036		
8	0.25	−0.018		
9	0.25	0.000		

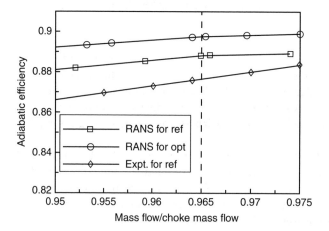

Figure 3.5 Adiabatic efficiency compared to mass flow rates. Source: Reproduced with permission from Samad et al. (2008a), (figure 5 from original source) Copyright © 2008 by the American Institute of Aeronautics and Astronautics, Inc.

3.1.4 Optimization Procedure

Single- and multi-objective optimization approaches are applied to numerical or experimental data. In surrogate-based optimization, surrogate models approximate the data
and predict approximate results. In multi-objective optimization, surrogate models are
coupled with genetic algorithm (GA), and optimum designs are predicted. Procedures of
surrogate-based optimization for single- and multi-objective optimizations are shown
in Figures 3.2 and 3.6, respectively.

3.1.5 Search Algorithm

An example of a search algorithm to find an optimum solution is the Sequential
Quadratic Programming (SQP) *fmincon* function in MATLAB (Mathworks Inc. 2016),
which is based on Newton's technique for constrained optimization. It is a three-step

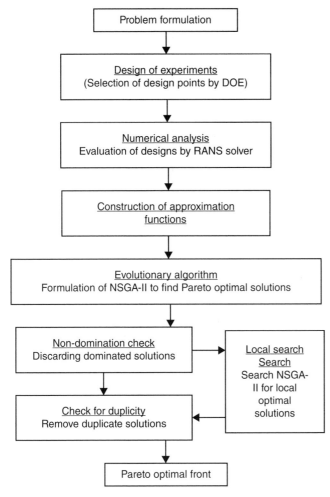

Figure 3.6 Multi-objective optimization procedure. Source: With kind permission from Springer Science+Business Media: Samad et al. (2008b), (figure 2 from original source), © 2008.

process; that is, update the Hessian matrix, solve for the quadratic programming, and form a search direction. The other search algorithms include GA, particle swarm optimization, and so on.

3.2   Multi-Objective Optimization (MOO)

MOO optimization deals with more than one objective function. The objective functions may be conflicting in nature and may have to maximize or minimize simultaneously. We deal with MOO on our day-to-day life. We minimize cost and maximize comfort. For example, while planning a tour we look for visiting maximum tourist spot at minimum cost and time. So, the objectives are conflicting. In practice, there can be more objectives that control our everyday life.

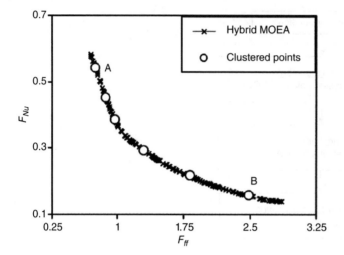

Figure 3.7 Pareto-optimal front. Source: With kind permission from Springer Science+Business Media: Samad et al. (2008b), (figure 4 from original source), © 2008.

For MOO, multiple solutions exist in a solution set. The results are presented as a Pareto-optimal Front (PoF) (Figure 3.7). The conflicting objectives produce several Pareto-optimal solutions. In MOO procedure, a simple weighted sum of objectives or Pareto front-based methods can be used. These approaches are explained next.

3.2.1 Weighted Sum Approach

The weighted sum of objective functions method (Collete and Siarry 2003) makes a MOO problem into a single-objective problem. Two objectives, F_1 and F_2, that are linearly joined by a weighting factor "w_f" form a single-objective "F_c" and the expression can be written as:

$$F_c = F_1 + w_f F_2 \tag{3.5}$$

The weighting factor (w_f) is selected based on the designer's own selection.

3.2.2 Pareto-Optimal Front

The MOO method gives a set of optimal solutions such that no other results are superior with respect to all the objectives. The solutions are represented in a PoF (Li and Padula 2004). MOO problems use GA or other algorithms to get a PoF (Li and Padula 2004; Queipo et al. 2005; Hwang and Masud 1979). The MOO method can be expressed as Figure 3.6:

$$\text{Minimize} \quad \overline{f}(\overline{x})$$

$$\text{Subject to} \quad \overline{g}(\overline{x}) \le 0$$

$$\overline{h}(\overline{x}) = 0 \tag{3.6}$$

The flow chart to produce a global PoF is presented in Figure 3.6. Constraints and objective functions are formulated mathematically and calculated on the data collected from experiments. An Italian engineer and economist Vilfredo Pareto (1848–1923) used the concept of optimality in his studies of economics. It was named Pareto-optimality.

The most favored solution can be obtained using various philosophies. MOO methods can be categorized into different classes such as a priori, a posteriori, interactive, and so on. All these classes contain preference information in diverse methods. In an a priori method, preference information is requested in advance and then a suitable solution satiating these preferences is found (Hwang and Masud 1979; Deb 2001). In an interactive method, the most desired solution is iteratively explored by the decision maker. The pareto-optimal solution is created in each iteration and the DM (design method) defines how the solution can be enhanced. The data provided by the DM is then considered while producing new pareto-optimal solution.

In NSGA-II (non-dominated sorting of genetic algorithm-II) (Deb 2001), parameters such as mutations, crossover, population size, and generation are adjusted to outfit the nature of the problem. The set of imprecise Pareto-optimal solutions achieved by NSGA-II is further processed to enhance the quality of Pareto-optimal solutions with a weighted sum scheme of local search methods. In this scheme, the objectives are united so it can be easier to implement. The weight for each objective is computed as:

$$\overline{w} = \frac{(F_j^{max} - F_j(X))/(F_j^{max} - F_j^{min})}{\sum_{k=1}^{M}(F_k^{max} - F_k(X))/(F_k^{max} - F_k^{mim})} \tag{3.7}$$

The combined objective becomes:

$$F_c = \sum_{k=1}^{M} F_k w_k \tag{3.8}$$

where, $\overline{w}$ and M are the weights for jth objective and the number of objectives, respectively. F_j^{min}, F_j^{max}, and $F_j(X)$ are the scaled minimum, the scaled maximum and the primary values of jth objectives, respectively. F_c is optimized by a search algorithm. The solutions are merged with the NSGA-II results and pareto-optimal solutions are obtained.

3.3 Constrained, Unconstrained, and Discrete Optimization

3.3.1 Constrained Optimization

As in the examples given at the beginning, we can buy items from our income, but there are several constraints to buy an expensive house or a diamond ring. We can also have the constraint of time. In engineering practice, we can have a higher efficiency compressor, but the system can develop higher stress concentration on the blade. The designer should limit the level of stress constriction by selecting the proper material and proper design at minimum possible cost.

In constrained optimization, an objective function can be optimized with respect to any variables in the presence of constraints on the variables. The constraints can be either hard or soft constraints. The conditions required to be satisfied for the variables are hard constraints. The variables are penalized in the objective function for the soft constraints if the conditions on the variables are not satisfied. Equation (3.6) shows the constrained optimization definition.

3.3.2 Unconstrained Optimization

Unconstrained optimization problems maximize or minimize an objective function depending on real variables without any restriction on their values (Unconstrained Optimization (n.d.)). Hence, the constraints in the Eq. (3.6) are not valid. Sometimes, to make it easier to solve an unconstrained problem, the constrained variable or objective is penalized to make it unconstrained. For example, in the problem mentioned in Section 3.1.3 (compressor problem), if the pressure ratio is constrained as 2.2 and is optimized for efficiency, the problem is constrained.

3.3.3 Discrete Optimization

In discrete optimization, some or all of the variables used in a discrete mathematical program are restricted to be discrete variables. A discrete combinatorial optimization has been used by Mosevich (1986) to design a hydraulic turbine runner.

3.4 Surrogate Modeling

3.4.1 Overview

"A surrogate" implies a borrowed concept or "something that replaces or used instead of something else" (Vale et al. 1996). The surrogates mimic the actual system and produce an approximate result in the engineering application. The CFD or FEM models are high-fidelity models and have more realistic function in experiments and therefore are called a "hi-fidelity" model. The surrogates approximate the CFD or FEM results where fidelity is low and is called a low-fidelity model.

The surrogate models basically take the design variable values and an objective function value to create approximate functions. There are numerous surrogates, such as response surface approximation (RSA), Kriging (KRG), artificial neural network (ANN), and support vector machine (SVM). The approximate functions produce different fitting and final design may not be optimal by the surrogates. The nature of the data points gives fitness. A surrogate may perform well for problem *A* but may not perform well for problem *B*. Again, the number of data points and sampling strategy also controls the accuracy or fitness of the surrogates.

3.4.2 Optimization Procedure

In a single-objective optimization approach, surrogate approximation models are implemented. Three basic surrogate models (RSA, KRG, and RBN) and a weighted-average surrogate (WAS) model are explained next. There are other WAS modeling techniques (Viana et al. 2014), and these are not discussed in this book.

In MOO, the surrogate models are coupled with GA and optimal designs are predicted.

3.4.3 Surrogate Modeling Approach

Almost all engineering design problems need simulations and experiments to assess objectives and constraints as the function of design variables. In real-world problems, a single simulation can get expensive in terms of time and cost. To reduce such expenses the use of surrogate models, which approximate the data points, can be applied. Such a model uses a limited number of data points. The challenge in surrogate modeling is to generate surrogates that have higher accuracy, while using a restricted number of sampling points. The surrogate construction process includes three steps:

- Selection of samples
- Construction of surrogate models and optimization of the model parameters
- Error analysis of the optimal design.

The accuracy of the surrogate model is subject to the quantity and the location of samples in the design space. Different design of experiment techniques produces errors because of noise in the data or errors occurring because of an incorrect surrogate model. The most commonly used surrogate models are polynomial RSA (Myers et al. 2016), KRG (Martin and Simpson 2005; Sacks et al. 1989), vector SVM (Cristianini and Shawe-Taylor 2000), and ANN (Orr 1996). These models are described in detail next.

3.4.3.1 Response Surface Approximation (RSA) Model

In RSA, the discrete responses achieved from numerical calculations produce a polynomial response function. This finds a relationship between design variables and response function. The linear response function can be expressed as:

$$F_i = \sum_{j=1}^{N} c_j x_{ij} + \varepsilon_i \tag{3.9}$$

where n is number design variables, x_j is vector of variables, and ε_i represents errors.

$$E(\varepsilon_i) = 0 \text{ and } V(\varepsilon_i) = \sigma^2 \tag{3.10}$$

σ^2: variance.

Equations (3.3) and (3.4) can be expressed in the matrix form as:

$$F = X \beta + \varepsilon \tag{3.11}$$

X is an $M{\times}n$ matrix of design variable values.

The least square estimate of β is

$$\hat{c} = (X^T X)^{-1} X^T F \tag{3.12}$$

The constructed second order polynomial response can be stated as:

$$F(x) = c_0 + \sum_{j=1}^{N} c_j x_j + \sum_{j=1}^{N} c_{jj} x_j^2 + \sum \sum_{i \neq j} c_{ij} x_i x_j \tag{3.13}$$

The number of regression coefficients is calculated by $\frac{(n+1)(n+2)}{2!}$ for a second-order polynomial model. The sum of squares of the residuals gives the fitness of the data into a curve and is formulated as

$$R^2{}_{adj} = 1 - \left(\frac{n-1}{n-p}(1 - R^2) \right) \tag{3.14}$$

$$R^2 = 1 - \frac{SS_E}{S_{yy}} \tag{3.15}$$

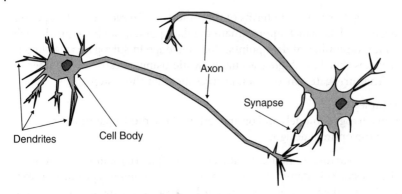

Figure 3.8 A neuron of the human body.

where, $SS_E = \sum_{i=1}^{n} (F_i - \hat{F}_i)^2$, $Syy = F'F - \dfrac{\left(\sum_{i=1}^{n} F_i\right)^2}{n}$ and p is the number of model parameters. This is specified by the number of regression coefficients.

3.4.3.2 Artificial Neural Network (ANN) Model

The ANN system in mathematical system works in a similar fashion to the nervous system works in the human body (Figure 3.8). We learn from the experiences, and once new information is absorbed by the sensors located in different part of the body, we respond. One example is a highly sensitive nasal sensor or receptors in the nasal passage that creates a sneeze, while a low sensitivity nerve may not respond well, even during an insect bite. Therefore, there is certain range where our sensory nerves work well. The nerves are trained, and the data are stored and analyzed in our brain. Similarly, the computational ANN model initially takes data (samples), trains itself, and decides their weights and biases. While a new piece of data is given as an input, the network (ANN) compares it with the existing data by the trained network and gives an output or the predicted value of the objective function. The ANN has a large number of parallel computational systems (simple processors) linked by adjustable interconnections. The concept in ANN is to mimic human functions such as predicting from earlier data, learning from experience, and so on.

The radial basis neural network (RBNN) (Orr 1996) has two layers: a hidden layer of radial basis and a transfer function with a linear output (Figure 3.9). The hidden layer

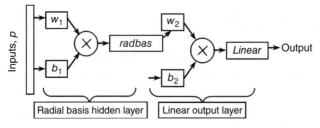

Figure 3.9 Radial basis network (single neuron). Source: Reproduced with permission from Samad and Kim (2009), (figure 2 from original source) © 2009 Turbomachinery Society of Japan, Korean Fluid Machinery Association, Chinese Society of Engineering Thermophysics, IAHR.

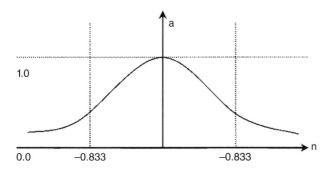

Figure 3.10 Radial basis function.

comprises a set of radial basis functions that act as activation functions, the reaction of which fluctuates with the distance between the center and the input.

The radial basis function (Figure 3.10) takes a value = 1 for input = 0. The output increases with a decrease in distance between w (weight) and p (input vector). The b (bias) allows adjusting the sensitivity of the *radbas* neuron. The design parameters are spread constant (SC) and a user-defined Error Goal (EG). The SC value is chosen carefully so that it should not create an insensitive network to all inputs or over sensitive to the inputs. The mean square EG selection is also vital. A very small EG leads to network over-training, while a large EG influences the model's accuracy.

3.4.3.3 Kriging Model (KRG) Model

The KRG model was first developed by a South African Engineer, D. G. Krige in 1960, and used for his mining application. Later, the model was adopted for design optimization purposes. In this model, the basic formulation comes from the input responses and fitting the curve based on regression analysis. The response function contains a systematic departure and a global model. Mathematically, we can express this as,

$$F(x) = f(x) + Z(x) \tag{3.16}$$

where the known function *f(x)* represents the trend over the design space and the unknown function *F(x)* is to be estimated. The function $Z(x)$ generates a localized deviation to interpolate the sampled data points with a Gaussian connection having zero mean and nonzero covariance. The covariance matrix of $Z(x)$ is expressed as

$$\text{cov}[Z(x^i), Z(x^j)] = \sigma^2 R[R(x^i, x^j)], \quad i, j = 1, 2. \ldots \ldots n_s \tag{3.17}$$

where R is a correlation matrix with a spatial correlation function $R(x^i, x^j)$ as its elements. The process variance σ^2 represents the spatial correlation function scalar for two sampled data points x^i and x^j. The Gaussian function is preferable for a gradient centered optimization algorithm.

3.4.3.4 PRESS-Based-Averaging (PBA) Model

As stated earlier, the WAS model makes an average by assigning weights to the surrogates and produces a global model (Goel et al. 2007; Viana et al. 2014). There are several weighted-average models that have been tried to unify the surrogates. The WTA3 model developed by Goel et al. (2007) was renamed the PBA model by Samad et al. (2008a, b). The predicted response for the PBA model is given by:

$$F_{PBA}(x) = \sum_{i}^{N_{SM}} w_i(x)F_i(x)$$ (3.18)

where, N_{SM} is the number of surrogate models that are used to build PBA. The ith surrogate model $(F_i(x))$ at design point x creates the weight $w_i(x)$.

The surrogate model that generates the highest error gets the lowest weight. The mean square error (MSE) or predicted error sum of squares (PRESS) gives the global weights. The weighting scheme is expressed as:

$$w_i^* = \left(\frac{E_i}{E_{avg}} + \xi \right)^{\kappa}, w_i = \frac{w_i^*}{\sum_i w_i^*}$$ (3.19)

$$E_{avg} = \sum_{i=1}^{N_{SM}} \frac{E_i}{N_{SM}}; \kappa < 0, \xi < 1$$

$$E_i = \sqrt{MSE_i}, i = 1, 2, \dots, N_{SM}$$

where $\xi = 0.05$ and $\kappa = -1$ (Goel et al. 2007). The constants ξ and κ are found through analytical function testing and iterations.

For the WAS method, the MSE calculations provide the weights. A cross-validation technique is implemented to calculate the MSE. The data produced through the CFD or other methods are separated into "k" subsets of approx. equal size and a surrogate is trained k times. One of the subsets is left out each time, and $k - 1$ data points construct the surrogate. The omitted subset is fed into the surrogate to get its response or objective function value. Hence, the difference between the value of the existing objective function and the predicted objective function value is the error for the omitted data. Now, the data points produce k surrogates and k errors. The k errors are calculated through mean square error estimation. The equation represents the MSE calculation:

$$MSE = \frac{1}{k} \sum_{i=1}^{k} (F_i - F_i^{i-1})^2$$ (3.20)

where $F_i^{(i-1)}$ is the calculation at $\mathbf{x}^{(i)}$ using the surrogate created with all sample points excluding $(\mathbf{x}^{(i)}, F_i)$.

3.4.3.5 Simple Average (SA) Model

An SA surrogate model is constructed when all the basis surrogates contribute equally to find the sensitivity of the weights calculated for the PBA model. Hence, the weights of the surrogates are computed using the formula:

$$w_i = \frac{1}{N_{SM}}$$ (3.21)

Equation (3.18) is used to formulate this surrogate. There are many other surrogate models including the variants of surrogates (Samad et al. 2008a, b; Forrester and Keane 2009). For further study, readers can go through the references (Forrester and Keane 2009).

3.5   Error Estimation

3.5.1   General Errors When Simulating and Optimizing a Turbomachinery System

The fitting surrogates include an error that comes from the data and may have an uncertainty or an approximation. The increasing computational power has helped in reducing errors in computational works as we can compute several terms of a Taylor series or include far more complex equations. This also allows us to include the uncertainty information in system analysis. To obtain a good result, the error should be minimal in our CFD, FEM modeling, or experimental data.

Error in design. Error in design occurs if any design used has an inherently erroneous design. Obviously, this will create problems during initial problem setup and final optimal design validation. For example, if a designer selects an infeasible design that might give an apparent result, it may not produce all results that are reliable.

Design by inexperienced designer. The other source of error can come from the meshing, discretization, boundary setting, turbulence model selection, improper domain selection, and so on. This error is negligible if the designer is experienced in CFD or FEM modeling. One obvious problem can be from the domain selection. A designer can define wind turbine blade boundary condition as far field to be two times of the blade diameter. In the same domain, if the turbine diameter is increased, the domain size may not be sufficient. Also, a distorted turbine blade requires proper meshing in each design under evaluation.

Error in post-processing. A designer can obtain reliable results from simulations but defining the output functions can be erroneous. Therefore, it should be checked carefully whether there is any error in post process.

Properly defined variables and objective functions are required. If a variable has less impact on an objective function or highly fluctuates due to uncertainty, the results may look incorrect or it may actually produce a wrong result.

Proper lower bound (LB) and upper bound (UB) are required. A LB or limit of a variable or an UB of a variable defines the design space. If too large a design space is selected, the included error may be higher. For instance, the optimal point may be far away from any of the limits or very near to a limit. The optimal point very close to the boundary may not have enough sample points to give gradient information. Alternatively, the same number of samples in a large design space may give poor gradient information near the optimal zone.

Enough and well-distributed data should be taken. For a specific design space, if the number of data is increased, then the un-sampled location is decreased. Figure 3.11 shows the density of design points in the design space. The top figure indicates that the density is low while the bottom figure shows the density is high. Hence the optimal points are also different.

A proper surrogate should be selected and modeled properly. For example, in an ANN model, the EG and SC should be selected properly or else a highly sensitive network or insensitive network is obtained. Too low or too high a value creates a non-conformal network. For a new input, the network may behave incorrectly. The sample location, sampling type, and design space size also matter.

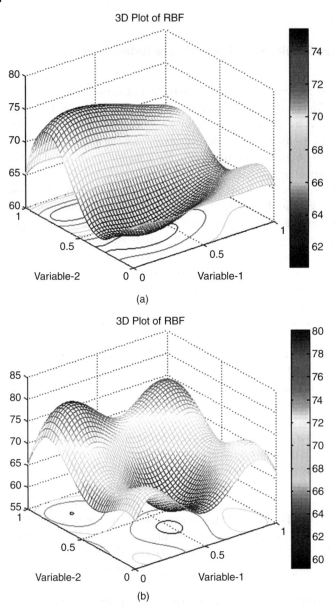

Figure 3.11 The un-sample location changes the optimal point (the z-axis shows the efficiency of a pump). (a) Optimal point using 16 sample spaces with the RBF surrogate. (b) Optimal point using 13 sample spaces with RBF surrogate.

Figure 3.12 shows that at certain values of SC and EG, the optimal point is changing. The right-hand side, the optimal point, shows a better result.

Similarly, the R_{2adj} parameter is important in the regression analysis. A designer should take care of such values. Normally, R_{2adj} near 1 (less than 1) is acceptable. Too small a value may give more of an approximation error when producing an optimal

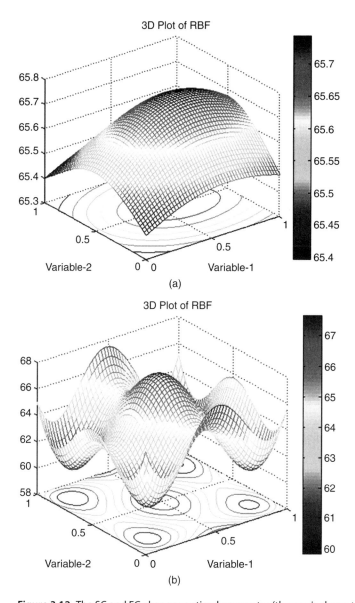

Figure 3.12 The SC and EG changes optimal parameter (the z-axis shows the efficiency of a pump). (a) EG = 0.9, SC = 0.51. (b) EG = 1.0, SC = 0.4.

design. To get high R_{2adj}, one should have lower noisy response data. At this stage, the designer should able to identify the noisy data and eliminate if required.

Table 3.4 shows that the initial 16 data points gave $R_{2adj} = 0.492$ and while removing some noisy data it reached $R_{2adj} = 0.995$. Here, removing too many data might also remove the non-noisy data. The removal is based on PRESS error estimation and removing the data can make the surface smooth. Hence, identification and removal are a part of the optimization process. Figure 3.13 shows the change in response surface due to noisy

Table 3.4 Effect of noisy data removal on R_{2adj}.

No. of samples	R_{2Adj}
16	0.492
15	0.761
14	0.823
13	0.87
12	0.903
11	0.954
10	0.986
9	0.995

data removal. In this case, too many data were removed sequentially and the optimal point location was also changed.

3.5.2 Error Estimation in Surrogate Modeling

The surrogate models are approximation models and the output response depends on the training of the surrogates. As stated earlier in Section 3.3, the surrogates are checked for their global error production capability. The cross-validation (CV) error gives an idea of error production. In a CV error, a set of data is taken and divided into k subset. Each time, $k - 1$ set data are used to train a surrogate and the remaining set is used to validate. Take an example given in Table 3.5.

It is assumed that there are two variables and three-level full factorial designs that are used to generate the sample points. The design space is defined by the ranges:

Variable 1, V1 = [−10 10]
Variable 2, V2 = [50 70]

The objective function values are obtained and presented in the last column after CFD simulation.

Step 1:

In CV, the first data is removed from the table (design number 1) and a table of eight sample points is taken (Table 3.6).
A surrogate model is constructed as follows:
Objective function,

$$Y = b_0 + b_1 x_1 + b_2 x_2 + b_{12} x_1 x_2 + b_{11} x_1^2 + b_{22} x_2^2 \tag{3.22}$$

Then, the design variables [10 50] are taken from design 1 and the value is fed to Eq. (3.22).
And, the value Y1 = f1′ is obtained. The objective function value in design number one is f1 = 72.
And, the error e1 = f1 − f1′

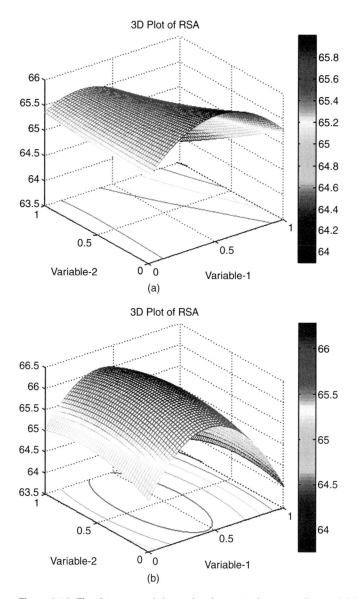

Figure 3.13 The data removal shows the change in the optimal point. (a) Sixteen data points are used. (b) Nine data points are used.

Step 2:

Next, the second design is removed from the table as shown in Table 3.7.

Hence, eight designs are taken as the second data is removed. Again, the error is calculated as in step 1:

$$e2 = f2 - f3'$$

Table 3.5 Design table.

Design number	Variable 1	Variable 2	Objective function
1	10	50	70
2	10	60	72
3	10	70	73
4	0	50	78
5	0	60	80
6	0	70	75
7	−10	50	74
8	−10	60	77
9	−10	70	72

Table 3.6 Design table: after the first design was removed.

Design number	Variable 1	Variable 2	Objective function values
2	10	60	72
3	10	70	73
4	0	50	78
5	0	60	80
6	0	70	75
7	−10	50	74
8	−10	60	77
9	−10	70	72

Table 3.7 Design table after the second design is eliminated.

Design number	Variable 1	Variable 2	Objective function values
1	10	50	70
3	10	70	73
4	0	50	78
5	0	60	80
6	0	70	75
7	−10	50	74
8	−10	60	77
9	−10	70	72

Step 3 to Step 9:

Following the same procedure given in steps 1 and 2, the errors e3, e4, ..., e9 are calculated.

In this manner, the errors are obtained for all the design points: $\text{rmse\_sur1} = \sqrt{\left(\sum e_i \right)}$. By following the same procedure for other surrogates, rmse_sur2, rmse_sur3, and so on are obtained. Comparing the surrogates gives the idea of the fitness of the surrogates.

For weight assignment, the formula given in Section 3.4.4.4 is used. The weights are based on this concept.

3.5.3 Sensitivity Analysis

The accuracy and effectiveness of a design and optimization depend on the factors or parameters under investigation. There are designs that need severe modification. For example, any design coming directly from any invention or from a concept design needs initial evaluation for optimization. These designs may not require complex mathematics at the initial stage but a simple analytical formulation or the experience of a designer. There are many parameters that can affect the system performance of a system severely. After the initial phase, the design is improved. At this level, the application of a complex optimization approach may not be useful because the initial result can direct the design path. Once this preliminary step is over, the interaction of multiple parameters affecting the performance is studied. For example, after the initial design of a turbine used for ocean energy harvesting is completed, the next step is to consider multiple parameters interacting with each other. Any change in one parameter affects another parameter. Design optimization based on experienced researchers without applying any optimization technique is long gone. Therefore, without any systematic optimization technique, performance improvement is difficult.

On the other hand, the turbomachinery systems that evolved several decades ago are already optimized. In the gas turbine system, increasing efficiency by 1% or a reduction in fuel consumption is a great achievement today. Therefore, researchers are trying not only to reduce fuel consumption, but they are also considering other parameters, such as structural stability, operating or surge related parameters, and application in difficult conditions such as extreme weather.

Many engineering designs pass through several design steps: preliminary design, detailed design, and design optimization. The task, especially in complex systems such as a gas turbine system is highly complex. It requires hundreds of engineers and technicians to work together. To go for a design optimization, one should have knowledge of all the aspects of the system they are trying to optimize.

3.5.3.1 Number of Variables and Performance Improvement

The effectiveness of surrogates is determined by the nature of the designs under consideration. The number of design points is governed by the number of design variables considered. If RSA is used, a number of design points required for n number of variables is $((n + 1)(n + 2)/2) + 1$. As a result, the number of design points increases with the increase in design variables.

Now the question is, how to eliminate the variables? A turbine can have variables such as

- Stacking
- Camber profile
- Leading edge (LE) and trailing edge (TE) shapes
- Chord length, number of blades, hub cap, rotor-stator interaction
- Number of stages
- Flow velocity, fluid property, temperature, turbine speed, and so on
- Blade height
- Blade sweeps, blade lean, and so on.

Objective functions: weight of the whole turbine system, efficiency, temperature ratio, pressure ration, entropy generation, stress, cost, operating range, manufacturing feasibility, and so on. Hence, any one or many objectives can be selected and optimized. It is easier to go for a single-objective optimization but multiple objectives can give a better design. Accordingly, the designer should select proper objective functions that can fulfill his/her needs.

3.5.3.2 Example of Sensitivity Analysis

In the compressor problem (Section 3.1.3), there are variables such as lean, sweep, and skew of a blade. Now, a designer can identify most important variable by bringing in a small change in the variable that brings the most significant change in performance. Figure 3.14 shows that sweep is more important than the variables lean or skew. Hence, a designer may ignore lean as its effect is negligible.

The figure is obtained from a surrogate. The efficiency maximized design was considered the base design and the other variable changes from +10% to −10%. The y-axis shows the change in efficiency, which is a ratio of the instantaneous value of efficiency minus the optimal efficiency divided by optimal efficiency.

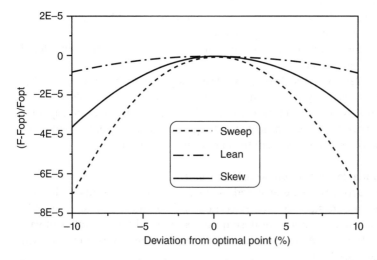

Figure 3.14 Sensitivity analyses for optimum shape by a surrogate.

3.6 Sampling Technique

3.6.1 Sampling

Sampling is a technique used to create a representative set of data that defines the entire population. To get the right picture, it is necessary to take random samples using a well-defined technique. Improper sampling will lead to a bias toward one set of opinions. One common example that can be found is television reporters who collect political opinions of a state undergoing elections based on opinions of people from different parts of the state at different age groups and professions.

3.6.2 Sample Size

The total number of samples required to make an average conclusion is also important. For example, it is necessary to include all age groups and various sections of the society of a country in order to make a conclusion about the average internet usage of the people. Therefore, an ample sample size is required to make the conclusions correctly. Otherwise, it may lead to inaccurate conclusions which are not representative of the entire population.

3.6.3 Design Space

In this survey, if the reporter takes samples only from cities and ignores rural areas that still do not have access to a proper internet connection, then the samples are still not sufficient enough to jump to a conclusion. This sample space is called design space.

3.6.4 Dimensionality Curse

In any CFD problem, simulating a turbomachinery system takes hours in the Reynolds-averaged Navier–Stokes (RANS)-based solver. Say that a compressor simulation takes 8 hours in normal desktop PC with a 4 GHz processor. If three-level full factorial is used for two design variables, the total number of designs will be nine and total simulation time will be 72 hours. If the number of variables increases, the simulation time increases accordingly. For a three-variable problem, it is expected to take $3^3 \times 8$ hours. So, for a 10-variable problem, it will take several months. Hence, reducing the number of design variables is an important part in creating samples.

3.6.5 Design of Experiments (DOE)

DOE is a structured and organized method to determine the association among factors influencing a process and its output. It is based on conducting and evaluating controlled tests to assess the factors that control the value of a parameter. DOE contains custom design, mixture design, response surface, screening design, Taguchi arrays, full factorial design, augment design, and so on (Forrester and Keane 2009).

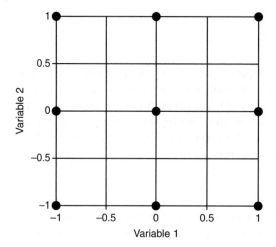

Figure 3.15 Three-level full factorial design.

3.6.6 Full Factorial Design

A full factorial design consists of numerous factors that can have several levels to take all possible combinations. Commonly, a two-level is used for the surrogate analysis. If two factors each with two levels are considered, it is usually called a 2×2 factorial design. For a larger number of levels, a fractional factorial design where some of the combinations are omitted can be considered. The full factorial design (Myers et al. 2016) contains all possible combinations of a set of factors with levels. A three-level full factorial design is presented in Figure 3.15.

3.6.7 Latin Hypercube Sampling (LHS)

LHS generates samples from a multidimensional distribution and is generally used for computer experiments (McKay et al. 2000). Assume a sampling function of N variables and the range of all variables are distributed into M equally possible intervals. The random samples are taken individually and it is important to remember which samples are reserved so far. The maximum number of combinations for LHS variables can be calculated by Eq. (3.23):

$$\left(\prod_{n=0}^{M-1} (M - n) \right)^{N-1} = (M!)^{N-1} \tag{3.23}$$

For instance, let's take a Latin hypercube of $M = 4$ divisions with $N = 2$ variables, it will have $(M!) N-1 = 24$ combinations. While a Latin hypercube of $M = 4$ divisions with $N = 3$ variables will give 576 combinations.

LHS generates random sample points and ensures a good point distribution in the design space. For two variables, LHS is presented in Figure 3.16. Figures 3.15 and 3.16 show the difference between the three-level full factorial and LHS designs.

Figure 3.16 Latin hyper cube sampling.

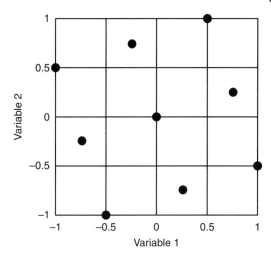

3.7 Optimizers

A typical optimization problem contains constraints, design parameters, variables, and objective functions. To solve such a problem, it is necessary to find a solver to evaluate the functions to find the minimum values satisfying the constraints. The optimizer selection based on the designer choice has wider options. Many optimizers are global point finders, while others are local point finders.

The options available to the designers (Documentation 2017a,b) are: linear, quadratic, least squares, smooth nonlinear, non-smooth, and so on. The local search methods include SQP, and simulated annealing methods, while an example of global search is GA.

The SQP method is used for mathematical problems where the constraints and the objective functions are differentiable. These are iterative approaches for nonlinear optimization. In the SQP method, the method reduces to Newton's method if the problem is unconstrained. The method is equivalent to Newton's method if the problem only has equality constraints of first order optimality conditions.

3.8 Multidisciplinary Design Optimization

3.8.1 What is Multidisciplinary Optimization?

The traditional engineering approach to design a complex system requires a team of people from multiple disciplines. For example, to develop a gas turbine system, structural engineers are required to check structural stability and vibrations. The aerodynamic experts will analyze fluid flow-related issues along with the shape of the turbine blade. The combustion engineers will study the combustion chamber performance and so on. These engineers aim to design an efficient system. The system should have reduced noise, higher stability, higher power output, higher operating range, lower material cost, lower weight, and so on.

The scenario has changed as the computer with higher specifications came into the picture. The CFD and FEM with optimization techniques have made tremendous developments in the aircraft industry. In the 1990s, the emergence of the supercomputer and parallel computers replaced traditional computers for research purposes. Commercial software has made life easier, although open source codes are also being used. Additionally, population-based algorithms have developed significantly.

Multidisciplinary design optimization (MDO) is an engineering field where numerous optimization methods are used to solve design problems integrating various disciplines. All related disciplines can be integrated simultaneously in MDO by a designer. The optimal combined-discipline problem can exploit the interfaces between the disciplines and is far better than designs obtained by optimizing each discipline in a sequence. However, the complexity of the problem increases with the inclusion of many disciplines simultaneously. The approaches are described briefly next.

3.8.2 Gradient-Based Methods

The structural optimization theory was developed by incorporating gradient-based methods and used optimality criteria along with scientific programming methods. The optimality criteria used the Karush–Kuhn–Tucker (KKT) environments for an optimized design. The conditions are used for structural problems like weight minimization with constraints, and so on. The mathematical programming used gradient-based methods. The gradient method of MDO is derived by combining optimality criteria with mathematical programming. The optimality criteria were good for displacement and stress constraints and the methodology helped to solve the twin problems of Lagrange multipliers.

3.8.3 Non-Gradient-Based Methods

The non-gradient-based MDO basically uses the GAs, simulated annealing, ant colony algorithms, and so on. Even when several models are available, researchers are still trying to find the best model for all applications. The code NSGA-II (Deb 2001) helps to get the Pareto front through the MDO approach.

3.8.4 Recent MDO Methods

The latest efforts are to take account of decomposition methods, evolutionary algorithms, approximation methods, response surface approach, and reliability-based optimization in MDO.

3.9 Inverse Design

3.9.1 Inverse Design versus Direct Design

The shape optimization can be done through two approaches: the inverse and direct design approach. In inverse design, it is necessary to specify the target parameters such as velocity or pressure (Shape Design Optimization (n.d.)). Their attainment greatly depends on the designer's expertise. The direct design approach can be categorized into global search methods and gradient-based methods. The global search methods approach a global optimum while the gradient-based methods approach a local optimum design. However, the cost of global search methods is higher with a huge number of design variables. The surrogate-based approach discussed in this chapter is a direct design approach.

3.9.2   Direct Design Optimization with CFD

The direct design method requires more computational time. Several commercial and open source codes are available that generate grids for the system automatically. One such software is the design Xplorer module in ANSYS workbench software that calls the functions automatically and the system gets optimized. There are still requirements for manual involvement to check accuracy and to monitor convergence for accurate results.

Because of complexity in flow through turbomachines, it is a difficult task to understand the flow completely and design the turbomachine. Any small change in blade geometry can bring about a massive change in performance in terms of head, efficiency, cavitation, flow separation, and vorticity. In industry, turbomachines are often designed based on a combination of designer's experience and direct CFD analyses of the flow inside the machines (Westra 2008).

3.9.3   Inverse Design Optimization with CFD

In a direct method, such as surrogate-based optimization method as described in this chapter, the input parameter is the geometry while output parameter is the flow field and its performance. However, the performance in terms of turbomachinery blade loading distribution, flow field, and blade curvature distribution are obtained through an inverse method. Here, the extra boundary condition is the blade loading function that is given for the inverse design method. The function may contain the information of mean-swirl distributions in the passage or velocity differences over blade surfaces. One simple version of the inverse design approach is given in Figure 3.17 (Kruyt and Westra 2014). The iterative approach contains a solution of the Laplace equation. The incompressible mass conservation law is given by,

$$\nabla^2 \phi = 0 \tag{3.24}$$

The blade geometry is changed accordingly by using the results acquired for the relative velocity at the blade surface (either at the suction side or the pressure side). This procedure is continued until it reaches convergence.

3.10   Automated Optimization

Automated optimization (AO) is defined as the method to integrate the computer-aided design (CAD) model, numerical analysis (fluid dynamics and/or structural mechanics), and optimization techniques in a single framework. To address CFD shape optimization, the designer must be able to alter the shape of several elements in the CAD model automatically. Then produce a grid of satisfactory quality for CFD simulations. The next step

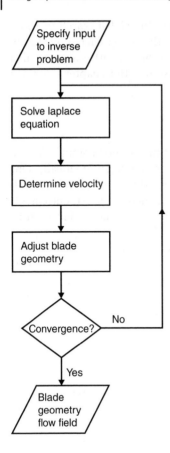

Figure 3.17 Flowchart of iterative scheme for solving inverse problems. Source: Reproduced with permission from Kruyt and Westra (2014), (figure 3 from original source) © 2014 IOP Publishing Ltd.

is to select an accurate and efficient CFD model to find the hydrodynamic performance of the system under study. The main parameter in selecting the modules of the automated CFD optimization is the ability to bridge these modules together in an automated manner, that is, without any user involvement. The AO method should also ensure that the fluid analysis is both accurate and efficient for the problem being modeled.

The basic process of AO is explained in Figure 3.18. To start the optimization procedure, first the design variables are selected. Selection of design variables is an important task in the whole procedure and must carried out only with a proper understanding of the problem. Only those design variables should be chosen that have the most effect on the performance of the system. The efficiency of the AO mainly depends on the number of design variables used for the model. Therefore, choosing unnecessary design variables might decrease the efficiency of the system. Once the CAD model is prepared, mesh generation is carried out followed by CFD simulation. With the help of numerical simulation, the constraints and objective functions are calculated and sent to the optimizer. The optimizer can use different optimization algorithms to find out the optima. Different case studies are mentioned later on the optimization algorithms used in literature. Once the result from the optimizer generates the optimum result, the process is terminated, or else different values of design variables are chosen and the whole cycle is run again until an optimal result is obtained.

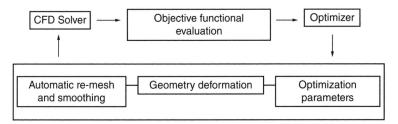

Figure 3.18 Flowchart of AO. Source: Reprinted from Javid Mahmoudzadeh Akherat et al. (2017), (figure 4 from original source) with permission from Elsevier.

The main difficulty in AO is to integrate the CAD software, CFD analysis software, and the optimization procedure. As the name suggests, AO requires the integration of CAD, CFD, and optimization in a single analysis loop without human intervention. Such integration of different software can be done using a programming language such as C++ or Python scripts.

3.10.1 Coupling Method with Adjoint CFD

AO is an area of research that is still quite new and there is a lot of research going on to develop a new methodology with regard to the coupling of different modules. Parametric modeling in the design process allows variation of geometry in an efficient way. For any simple to fairly complex geometry, an engineering process can have parameters in the range of 20–50. This means if we consider the direct integration of CFD and optimization of a complex geometry, there will be around 50 function evaluations (i.e., CFD computations) for each loop, which is computationally costly. As a result, direct coupling of CAD and CFD is not feasible when dealing with a complex geometry.

The adjoint method, instead of directly evaluating the change in objective function due to a variation in the parameter value, calculates the variation of the parameter value for the desired change in the objective function. Thus, the adjoint method works in a reverse way. Within a single computation, the adjoint method will yield the full gradient of the objective function, irrespective of the number of parameters. The full analysis would require a conventional (primal) CFD computation followed by one adjoint computation for each considered objective (AO using Adjoint Flow Solvers 2017). Adjoint CFD analysis for shape optimization is tested with turbine blades and airfoil shape optimization. It shows improvement in computational time compared to direct AO. AO using an adjoint flow solver is shown by the flowchart in Figure 3.19.

3.10.2 Case Studies

3.10.2.1 CFD-Based Design Automated Design Optimization for Hydro Turbines
A CFD-based design system with the combination of different blade design methods, automatic grid generator, and efficient optimization algorithms has been developed by Wu et al. (2007) The design system is applied to a rehabilitation project that provides 3% growth in efficiency and 13% improvement in power. The CFD-based design system developed by the authors is given in Figure 3.20.

The objective functions chosen for this problem are the head development, the mass flow rate, the rated power and the peak efficiency. The design variables are some chosen

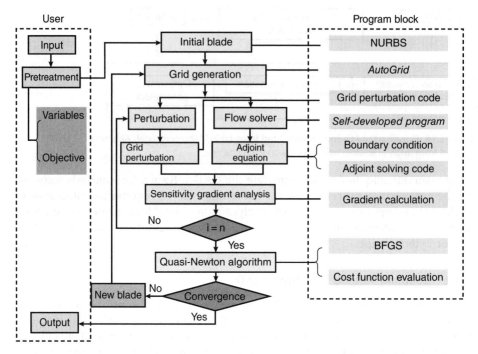

Figure 3.19 Flowchart of AO using adjoint solver. Source: Reprinted from Zhang et al. (2017), (figure 2 from original source) with permission from Elsevier.

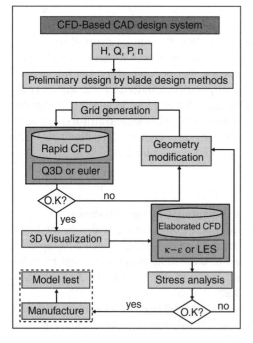

Figure 3.20 Schematic of AO. Source: Reproduced with permission from Wu et al. (2007), (figure 1 from original source) Copyright © 2007 by American Society of Mechanical Engineers.

points on the blade profile. Preliminary blade design is carried out using the conventional methods. After the grid generation, a CFD analysis is carried out for the same using Q3D CFD code, which is similar to any commercial CFD codes available. However, Q3D is used only for preliminary CFD analysis, as it takes less computational time. Once the optimization is carried out and the optimal geometry is reached, the CFD analysis is carried out using a commercial 3D Navier–Stokes code like ANSYS – FLUENT®. The optimum result shows improvement in terms of performance characteristics of the turbine.

3.10.2.2 AO with OPAL++

OPAL++ (OPtimization Algorithm Library ++) (Daróczy and Janiga 2016) is an in-house code of Otto von Guericke University Magdeburg, Germany developed to couple the CFD code and optimization algorithms. This is an object-oriented MOO and parameterization framework created in C++. This supports both Windows and Linux operating systems.

In order to create an optimization setup, OPAL++ requires two script files, the master and simulation script. The master script contains the definition of the optimization, the list of necessary input and output files and the settings for evolutionary algorithms. The second file, the simulation script, contains the workflow for evaluating a single design variable. OPAL++ provides many different commands to ease the coupling of different software. Both script files rely on "Language for OPAL++ Scripting" (LOS), a specific script language developed for the present application. Thanks to its very simple syntax, LOS has a very steep learning curve and eliminates the need to set up many different scripts in different languages (Daróczy and Janiga 2016). OPAL++ relies on parallel computing using Message Passing Interface (MPI). A master node controls the operations for selection of evolutionary algorithms. The master node starts the CFD solver and it can be run using many parallel CPU cores. Mohamed (2011) carried out AO of a Wells and Savonius turbine using the OPAL code. The schematic illustration of the process is shown in Figure 3.21.

For this work, commercial tool GAMBIT® is used for geometry creation and mesh development, and ANSYS – FLUENT® is used to simulate the flow field across the turbines. The objective function is calculated from the CFD results. The whole procedure is made automated by using journal files of GAMBIT and ANSYS – FLUENT. The main program is created in C, which calls the entire program sequentially as shown in Figure 3.20. The OPAL code itself decides which optimization code to use depending upon the CFD results. The optimization procedure is able to handle both single and MOO and the application results can be found in the literature.

3.10.2.3 PADRAM: Parametric Design and Rapid Meshing System for Turbomachinery Optimization

The main difficulty in AO is creating a good quality mesh automatically depending on the change in geometry. Much effort has been put together by the research community to come up with a reliable tool for the same. PADRAM is capable of optimizing multi-passage 3D blades and its circumferential pattern is carried out concurrently in one system (Shahpar and Lapworth 2003). This is built specially for turbomachinery applications, where the turbine geometry is defined at numerous radial stream sections. At the starting point, the PADRAM system generates the geometry and sends it for

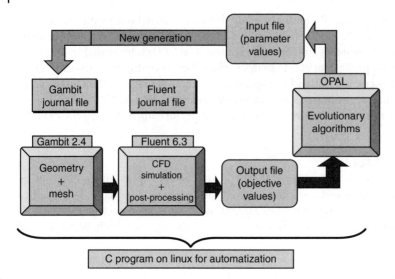

Figure 3.21 Schematic of AO using OPAL. Source: Reprinted from Mohamed et al. (2011), (figure 5 from original source) with permission from Elsevier.

mesh generation. PADRAM meshing begins by partitioning the computational domain into blocks and sub-blocks with the intention of generating algebraic grid and associated control functions. H-type meshes are used close to the periodic boundary and O-type meshes are given for the blades. Algebraically generated grids in each block are smoothed by solving a system of elliptical partial differential equations (PDEs) by the PADRAM (Shahpar and Lapworth 2003).

PADRAM requires a reference geometry to start the optimization procedure. The system is already equipped with different design parameters such as blade sweep or blade lean that are used to construct new optimized blade designs, see Figure 3.22.

The PADRAM system has the capability to work remotely to increase the computational power. The mesh generation can be viewed by the user, whereas other processes can run in the background. The system is fast and gives satisfying result both qualitatively and quantitatively. While PADRAM was initially built for turbomachinery application, over the years it has matured sufficiently. Figure 3.21 shows a range of gas turbine engine components where a PADRAM-based design can be used. Practically, the code can mesh all components of the latest gas turbine engine.

Rolls-Royce use the SOPHY (SOft-Padram-HYdra) system, a completely incorporated flexible aerodynamic design system designed based on PADRAM. SOPHY is used for modeling and optimization of turbomachinery mechanisms by making use of high-fidelity CFD codes. These modules are designed in such a way that they run in batch mode. Therefore, scripting and linking these modules are direct, which allows for efficient utilization of CFD for a fully automated and faster simulation. The optimization progression taking place in SOPHY is shown in Figure 3.23.

3.10.2.4 Problems of AO

The difficulties faced during CFD-based optimization, is summarized by Daróczy and Janiga (2016):

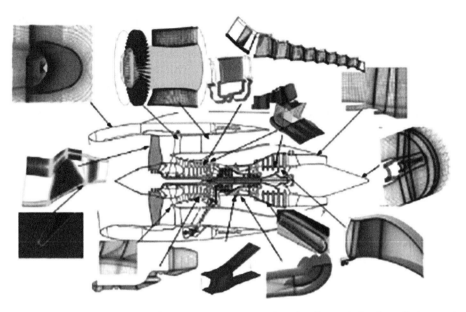

Figure 3.22 Application areas of PADRAM. Source: Reproduced with permission from Shahparand Lapworth (2003), (figure 1.1 from original source), Copyright © 2003 by American Society of Mechanical Engineers.

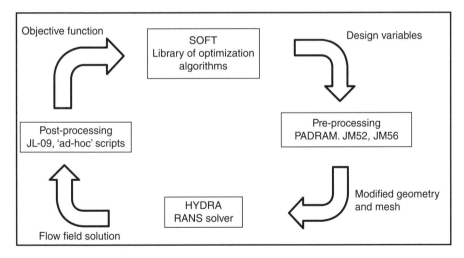

Figure 3.23 AO using the SOPHY system. Source: Reproduced with permission from Shahpar and Caloni (2013), (figure 1 from original source), Copyright © 2013 by American Society of Mechanical Engineers.

- Objective function values are not explicitly known and have to be computed based on numerical simulations. Thus, gradients are usually not available.
- Function evaluations are very costly, requiring from a couple of minutes of computational time up to several days.

- Due to numerical noise and model uncertainties, the objective functions are usually noisy.
- During the optimization, differences of the variables below manufacturing tolerance are irrelevant (e.g., optimization of a car geometry with nm precision).
- The geometry and mesh have to be created/morphed for each design variable in an automated and robust way.
- Different software (including proprietary commercial software) have to be coupled to cooperate for the optimization.

As a result, speed and efficiency are of key importance in AO.

3.11 Conclusions

This chapter discussed the optimization methodology used for optimization. There are ample methods, but only a limited number of methods were discussed. The discussion also included the AO approach along with a few case studies.

References

Automated Optimization using Adjoint Flow Solvers › CAESES. (2017). Retrieved June 02, 2017, from https://www.caeses.com/blog/2017/automated-optimization-using-adjoint-flow-solvers.

Collette, Y. and Siarry, P. (2003). *Multiobjective Optimization: Principles and Case Studies*. New York: Springer.

Cristianini, N. and Shawe-Taylor, J. (2000). *An Introduction to Support Vector Machines and Other Kernel-based Learning Methods*. Cambridge: Cambridge University Press.

Daróczy, L. and Janiga, G. (2016). *Practical Issues in the Optimization of CFD Based Engineering Problems*. Magdeburg: University Library.

Deb, K. (2001). *Multi-Objective Optimization Using Evolutionary Algorithms*. Wiley https://doi.org/10.1109/TEVC.2002.804322.

Documentation. (2017a). Retrieved June 02, 2017, from http://kr.mathworks.com/help/optim/ug/optimization-decision-table.html?requestedDomain=kr.mathworks.com#brhkghv-21.

Documentation. (2017b). Retrieved June 02, 2017, from http://kr.mathworks.com/help/optim/ug/choosing-a-solver.html#brhkghv-19, accessed on June 2, 2017.

Forrester, A.I.J. and Keane, A.J. (2009). Recent advances in surrogate-based optimization. *Progress in Aerospace Sciences* https://doi.org/10.1016/j.paerosci.2008.11.001.

Goel, T., Haftka, R.T., Shyy, W., and Queipo, N.V. (2007). Ensemble of surrogates. *Structural and Multidisciplinary Optimization* 33 (3): 199–216. https://doi.org/10.1007/s00158-006-0051-9.

Hwang, C.-L. and Masud, A.S.M. (1979). Methods for multiple objective decision making. In: *Multiple Objective Decision Making – Methods and Applications: A State-of-the-Art Survey*, Lecture Notes in Economics and Mathematical Systems, vol. 164, 21–283. Springer Science+Business Media https://doi.org/10.1007/978-3-642-45511-7.

Javid Mahmoudzadeh Akherat, S., Cassel, K., Boghosian, M. et al. (2017). A predictive framework to elucidate venous stenosis: CFD & shape optimization. *Computer Methods in Applied Mechanics and Engineering* 321: 46–69. https://doi.org/10.1016/j.cma.2017.03 .036.

Kruyt, N.P. and Westra, R.W. (2014). On the inverse problem of blade design for centrifugal pumps and fans. *Inverse Problems* 30 (6): https://doi.org/10.1088/0266-5611/30/6/ 065003.

Li, W., and Padula, S. (2004). Approximation methods for conceptual design of complex systems. In *Eleventh International Conference on Approximation Theory* (eds. Chui, C., Neaumtu, M., and Schumaker, L.). Gatlinburg, TN, May 2004.

Martin, J.D. and Simpson, T.W. (2005). Use of Kriging models to approximate deterministic computer models. *AIAA Journal* 43 (4): 853–863. https://doi.org/10.2514/1.8650.

McKay, M.D., Beckman, R.J., and Conover, W.J. (2000). A comparison of three methods for selecting values of input variables in the analysis of output from a computer code. *Technometrics* 42 (1): 55–61. https://doi.org/10.1080/00401706.2000.10485979.

Mohamed, M.H.A. (2011). *Design Optimization of Savonius and Wells Turbines*. Otto von-Guericke University Magdeburg.

Mohamed, M., Janiga, G., Pap, E., and Thévenin, D. (2011). Optimal blade shape of a modified Savonius turbine using an obstacle shielding the returning blade. *Energy Conversion and Management* 52 (1): 236–242.

Mosevich, J. (1986). Balancing hydraulic turbine runners – A discrete combinatorial optimization problem. *European Journal of Operational Research* 26 (2): 202–204. https://doi.org/10.1016/0377-2217(86)90181-5.

Myers, R. H., Montgomery, D. C., and Anderson-Cook, C. M. (2016). *Response Surface Methodology: Process and Product Optimization Using Designed Experiments*. Wiley Series in Probability and Statistics Established. https://doi.org/10.1017/ CBO9781107415324.004.

Orr, M. (1996). Introduction to radial basis function networks. *University of Edinburg*, 1–7. Retrieved from http://dns2.icar.cnr.it/manco/Teaching/2005/datamining/articoli/ RBFNetworks.pdf.

Queipo, N.V., Haftka, R.T., Shyy, W. et al. (2005). Surrogate-based analysis and optimization. *Progress in Aerospace Sciences* 41 (1): 1–28. https://doi.org/10.1016/j .paerosci.2005.02.001.

Sacks, J., Schiller, S.B., and Welch, W.J. (1989). Designs for computer experiments. *Technometrics* 31 (1): 41–47.

Samad, A. and Kim, K. (2008). Multi-objective optimization of an axial compressor blade. *Journal of Mechanical Science and Technology* 22 (5): 999–1007. https://doi.org/10.1007/ s12206-008-0122-5.

Samad, A. and Kim, K.Y. (2009). Surrogate based optimization techniques for aerodynamic design of turbomachinery. *International Journal of Fluid Machinery and Systems* 2 (2): 179–188.

Samad, A., Kim, K.-Y., Goel, T. et al. (2008a). Multiple surrogate modeling for axial compressor blade shape optimization. *Journal of Propulsion and Power* 24 (2): 301–310. https://doi.org/10.2514/1.28999.

Samad, A., Lee, K., and Kim, K. (2008b). Multi-objective optimization of a dimpled channel for heat transfer augmentation. *Heat and Mass Transfer* 45 (2): 207–217. https://doi.org/ 10.1007/s00231-008-0420-6.

Shahpar, S. and Caloni, S. (2013). Aerodynamic optimization of high-pressure turbines for lean-burn combustion system. *Journal of Engineering for Gas Turbines and Power* 135 (5): 055001.

Shahpar, S., and Lapworth, L. (2003). PADRAM: Parametric Design and Rapid Meshing System for Turbomachinery Optimisation. *ASME Conference Proceedings, 2003(36894),* 579–590. https://doi.org/10.1115/GT2003-38698.

Shape Design Optimization (n.d.). Retrieved June 02, 2017, from http://www.cfd-online .com/Wiki/Shape_Design_Optimization.

The Mathworks Inc. (2016). MATLAB – MathWorks. https://doi.org/2016-11-26.

Unconstrained Optimization (n.d.). Retrieved June 02, 2017, from http://www.neos-guide .org/content/unconstrained-optimization.

Vale, D., Mullaney, S., and Hartas, L. (1996). *The Cambridge Dictionary*. Cambridge: Cambridge University Press.

Viana, F.A.C., Simpson, T.W., Balabanov, V., and Toropov, V. (2014). Special section on multidisciplinary design optimization: metamodeling in multidisciplinary design optimization: how far have we really come? *AIAA Journal* 52 (4): 670–690. https://doi .org/10.2514/1.J052375.

Westra, R. W. (2008). Inverse-Design and Optimization Methods for Centrifugal Pump Impellers Enschede: Engineering Fluid Dynamics. Ph.D. thesis. https://doi.org/10.3990/1 .9789036527026. University of Twente.

Wu, J., Shimmei, K., Tani, K. et al. (2007). CFD-based design optimization for hydro turbines. *Journal of Fluids Engineering* 129 (2): 159. https://doi.org/10.1115/1.2409363.

Zhang, P., Lu, J., Song, L., and Feng, Z. (2017). Study on continuous adjoint optimization with turbulence models for aerodynamic performance and heat transfer in turbomachinery cascades. *International Journal of Heat and Mass Transfer* 104: 1069–1082. https://doi.org/10.1016/j.ijheatmasstransfer.2016.08.103.

4

Optimization of Industrial Fluid Machinery

Optimization techniques for maximizing industrial fluid machinery performance are described in this chapter. Both single- and multi-objective methods are explained in detail that are capable of dealing with both single- and/or multi-point problems.

The most common and relevant problem type concerns shape optimization of machine stationary and rotating parts. Regardless of the type of problem formulation, fluid machinery shape optimization is typically performed using an iterative computer loop that always involves at least three types of block: (i) an optimization engine (or tool) – as described in Chapter 3 – linked to (ii) a flow solver, and (iii) a parametric model of the machine shape. Whatever its type, the optimization engine drives the search toward optimal solutions and manipulates the decision variables defined within the parametric model. In a conventional optimization loop, each candidate shape produced by the parametric model is evaluated by means of a flow solver that, in turn, sends back quantitative information on the objective function/s value/s to the optimization engine. Such a process is iterated until an optimum set of solutions is found.

In the following, we shall implicitly refer to this loop architecture as the "baseline loop structure." Other, more sophisticated loop architectures have been widely adopted in recent years and will be described specifically when necessary.

Pump optimization is described first. Centrifugal pumps are dealt with as they are the most general case in terms of geometrical complexity followed by axial-flow and mixed-flow pump optimization. Following this, compressor and turbine optimization is described. Next, optimizations of fans, hydraulic turbines, and other types are discussed.

4.1 Pumps

4.1.1 Centrifugal, Mixed-Flow, and Axial-Flow Pumps

4.1.1.1 Centrifugal (or Radial) Pumps

Centrifugal pumps are widely used, either in single- or multistage configurations, in almost every plant involving circulation of liquids where a high head is required at a relatively low flow rate. Typical usages include pumping of water, petroleum, and petrochemical products. Other, more peculiar utilizations are found in water-based energy storage or in rocket propulsion systems, where pumps are used to pressurize gases at their liquid state.

Design Optimization of Fluid Machinery: Applying Computational Fluid Dynamics and Numerical Optimization, First Edition. Kwang-Yong Kim, Abdus Samad, and Ernesto Benini.
© 2019 John Wiley & Sons Singapore Pte. Ltd. Published 2019 by John Wiley & Sons Singapore Pte. Ltd.

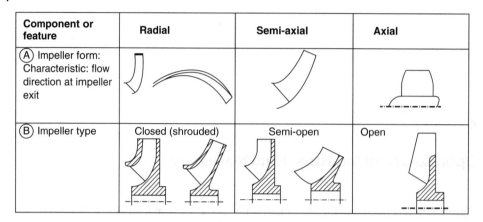

Figure 4.1 Types of pump impellers.

Nowadays, a centrifugal pump can reach up to 92% hydraulic efficiency in some large-scale, shrouded impeller-type (Figure 4.1) special applications at its Best Efficiency Point (BEP) in the pump map. However, figures less than 50% are not that uncommon in small units used for domestic pumping. This suggests that there is valid room for performance optimization in several pumping systems, which would lead to a dramatic reduction in the pumping power spent worldwide. Furthermore, optimum pump configuration is a trade-off among multiple and often conflicting objectives including reliability, low-cost manufacturing, and hydraulic efficiency.

Although centrifugal pumps are designed primarily for utilization close to their BEP, often they are operated far from this condition. In these circumstances, pumps operate at reduced efficiency and, at larger flow rates compared to BEP, they are prone to cavitation. Stall and pressure pulsations are typical of use at lower flow rates instead. Optimization of pump performance in fact might include several working points, including those at off-design.

4.1.1.2 Mixed-Flow and Axial-Flow Pumps

As the pump specific speed increases, that is, when the ratio of the head rise to the flow rate gets lower compared to centrifugal pumps for a given rotational speed, mixed-flow (or semi-axial in Figure 4.1) and axial-flow pumps are employed. These pumps are often found in water irrigation and drainage, as well as in several municipal and industrial pumping applications, like primary supplies in water networks, industrial processes, thermoelectric power plant condensing, and other circulating systems. A very particular application is devoted to high-speed waterjet propulsion of medium-to-large-scale marine vessels. Although mixed- and axial-flow pumps are designed especially for single-stage architectures, some examples are found of multistage machines when the head rise is typically greater than 60 m. In practice, the number of stages is limited to two or three. While axial-flow pump impellers are almost invariably open (i.e., unshrouded, Figure 4.1), semi-axial impellers often come in a semi-open configuration.

Experience shows that, in general, both mixed- and axial-flow pumps do have slightly smaller hydraulic efficiencies compared to centrifugal types. While maximum figures around 91% are documented at BEP for large-scale multi-megawatt mixed-flow machines, quite rarely an axial-flow pump can reach up to maximum 90% at its BEP.

Also in such machines, actual figures for smaller scale machines are much lower and often drop back to 40% in low-power installations. Operation at off-design conditions is also important for such machines and a close look at the entire operating range should be given and properly considered in the optimization.

The following is a fairly complete review of the most advanced optimization methods available today in both industry and academia for pump design optimization. Organizing and harmonizing such an abundant source of information is not an easy task. For this reason, we decided to present the most relevant optimization methods based on the type of aforementioned loop blocks: first, the parametric models will be dealt with along with the pertinent flow solver type, and a close-up on optimization engines will be given.

4.1.2 Parametric Shape Models and Flow Solvers for Pump Optimization

Several types of models have been used so far to describe and parameterize a centrifugal pump geometry including both rotating and stationary elements. Roughly speaking, this involves the choice of the number of geometrical dimensions that the model has to deal with; that is, from one to three. Based on this choice, a pertinent flow solver type is applied accordingly; that is, from a simple 1D to a more complex 3D one. Furthermore, since unsteady investigations can be sought, one option more (the one related to time) is available that can be regarded as a further dimension. As it has been widely discussed in the previous chapters, within the class of flow solvers available today several choices are possible such that, in general, for a given type of geometrical approach several examples of flow-solutions can be produced. Just as an example, a two-dimensional pump radial cascade can be analyzed using either the Navier–Stokes or Euler analyses depending on the choice on the objective function/s and on the computational resources available. Analogously, a spatial model of a pump impeller requires fully 3D flow analyses, which can be either of the RANS or DNS type.

In the following, according to the level of complexity required by the designer, we shall refer to 1D, 2D, and 3D geometrical and flow models.

4.1.2.1 1D Models

1D methods are still very useful for preliminary design, multi-point design, and product family design. In such cases, in fact, CFD techniques are too expensive for dealing with a vast exploration of the design space, especially when multiple, off-design evaluations are needed. 1D, meanline pump-flow on/off-design modeling methods have been developed in the past and are still today used for preliminary designs and performance analyses. Moreover, such models can be easily implemented and used as performance calculators that provide a rapid evaluation of candidate pump design concepts.

Examples of pump meanline model implementations are almost uncountable in open literature and the reader is referred to Japikse et al. (2006) for a complete review. Basically, they are divided into three levels of increasing complexity and accuracy.

A first step in design/analysis technique is known as the "scaling method," and is based on scaling of an existing design by means of the theory of similitude to produce a new candidate design. Dixon and Hall (2014), Whitfield and Baines (2002), and Japikse and Baines (1997) provide a description of methodologies where similitude laws are applied to the scaling of centrifugal machines. A second-step analysis follows the first step analysis and combines performance correlations, such as rotor efficiency, to provide a prediction of machine performance. Rodgers (1980) gives a detailed example of a second-step analysis using component correlations. The second-step design allows for greater flexibility than a first step, but it is somehow limited by the range and accuracy of the component performance correlations used in the model and the need for model calibration. Third-step analyses are based on a comprehensive set of fluid dynamic models to predict machine performance. Since in a third-step analysis fluid dynamic models are implemented, new machines can be theoretically developed that deviate from the realm of past experience. Relevant examples of such models have been recently described in Pelton (2007), where empirical models to predict pump model coefficients were described based on test results for a large dataset of pump and compressor impellers equipped with a vaneless diffuser followed by a volute. A similar procedure has also been outlined in Bitter (2007) for vaned diffusers. In two studies, an enhanced TEIS (Two Element In Series) and two-zone flow models (see Japikse 1996, 2001) were used for pump performance prediction, and regression models were developed using weighting factors so that a better matching between predicted and experimental data was achieved compared to the original work (Japikse 2001), where resulting models did not yield results that could be used with confidence in practice. In spite of the remarkable improvements obtained in Pelton (2007), continued effort needs to be spent to add further cases to datasets available and further validation is needed. As a matter of fact, these level three models include so many unknown parameters to be determined either by regression or other statistical procedures that they cannot be efficiently used unless a very large amount of high-quality experimental data is available, a condition that in fact prevents these methods from being applicable in optimization problems.

To the authors' option, the so-called second-step analysis constitutes a satisfactory trade-off between accuracy and simplicity. These are features that make it possible to reasonably calibrate a meanline model using a limited amount of experimental data. Instead of using parametric statistic methods, an optimization technique can be effectively used for proper calibration.

In the following, a level-two meanline prediction method suitable for performance prediction of turbopumps is proposed. Such a method is able to predict both design and off-design performance of a pump given the loss of the diffusion system at the design point. A limited number of free parameters are introduced in the model, which eventually makes it possible to calibrate it on real pump performance results using an optimization technique. In such models, typical decision variables include the rotor flow path radii, span, and blade angles at both the leading and trailing edges (Figure 4.2). The diffusion system variables include inlet and exit radii of vaneless section, throat areas in volute, or diffuser, leading edge angle of volute tongue or vaned diffuser (if present), channel spans, and stage exit area. Inlet flow conditions are bounded at the design point in terms of fluid flow rate, pump design rotational speed, pressure, temperature, and inlet flow swirl (if present). In a multistage pump, inlet conditions of the next stage are calculated and provided to the model from the outlet conditions of the previous stage.

4.1.2.1.1 Meanline Turbopump Model

A meanline pump-flow modeling method is developed as proposed in Veres (1994) that can predict the performance of pumps (head, torque, power efficiency and Net Positive Suction Head (NPSH)) at on/off-design operating conditions by providing a

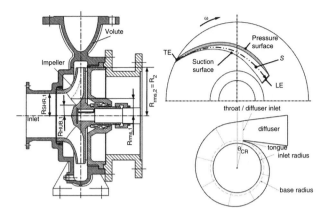

Figure 4.2 Decision variables in a meanline pump impeller and volute.

loss in the diffusion system at design conditions. Impeller efficiency and flow slip factor at the design point are given from empirical correlations provided in Veres (1994) to rotor-specific speed and geometry; however, these correlations are changed here to account for size effects and best approximate experimental data at our disposal. It is worth recalling that the procedure outlined in Veres (1994) relates chiefly to turbopumps used in aerospace applications, and therefore the results obtained hereafter should be rigorously applicable to this particular case. However, it has been demonstrated that a proper model calibration can make it possible to extend the validity of the model to industrial applications (Benini and Cenzon 2009).

The model is based on the Euler equation for determining the work exchange at the mid radius of pump runner (see Figure 4.2) coupled with empirical correlations for hydraulic rotor efficiency and slip factor. After establishing design point performance, off-design characteristic map can be estimated based on loss correlations. The loss in the diffusion system at the design point is given and then changed at off-design by means of an empirical relation. The result of the matching between the pump impeller and the diffusion system determines the pump performance chart and influences the location of the stall and cavitation inception lines.

Referring to the nomenclature and to Figure 4.2, in the following subsections are the equations used in the model.

4.1.2.1.2 Design Operation – Rotor Inlet

Flow area at the rotor inlet:

$$A_1 = \pi(R_{SHR,1}^2 - R_{HUB,1}^2) - B_{k,1} \tag{4.1}$$

where

$$B_{k,1} = Th_1 \cdot B_1 \cdot Z_1 / \sin \beta_{B,1} \tag{4.2}$$

Th is the blade LE thickness, Z is the number of primary blades and B_1 is the blade metal angle at the LE. For twisted blades, the model considers a parabolic variation of the blade angle along the leading edge and attributes a root-mean-square blade angle to the blade inlet.

Meridional, tangential components of flow velocity, as well as absolute flow velocity magnitude at impeller leading edge mean diameter are given as:

$$C_{M,1} = \frac{\dot{m}}{\rho \cdot A_1}, \quad C_{U,1} = \frac{C_{M,1}}{\tan \alpha_1}, \quad C_1 = \sqrt{C_{M,1}{}^2 + C_{U,1}{}^2} \tag{4.3}$$

Relative flow angle at rotor inlet are as follows:

$$\beta_{F,1} = \arctan \left(\frac{C_{M,1}}{W_{U,1}} \right) = \arctan \left(\frac{C_{M,1}}{U_1 - C_{U,1}} \right) \tag{4.4}$$

where $U_1 = 2\pi R_{rms,1} N$ is the blade tangential velocity.

Incidence angle (i.e., difference between blade angle and relative flow angle):

$$i_1 = \beta_{B,1} - \beta_{F,1} \tag{4.5}$$

Tangential component of flow relative velocity and relative flow velocity magnitude:

$$W_{U,1} = U_1 - C_{U,1} = C_{M,1} \cdot \tan(\pi/2 - \beta_{F,1}) \tag{4.6}$$

$$W_1 = \sqrt{W_{M,1}{}^2 + W_{U,1}{}^2} \tag{4.7}$$

$$C_{U,2} = U_2 + W_{U,2} \quad W_{M,1} = C_{M,1}$$

4.1.2.1.3   Design Operation – Rotor Outlet

Flow area at the rotor trailing edge:

$$A_2 = 2\pi B_2 \cdot R_{rms,2} - B_{k,2} \tag{4.8}$$

$$B_{k,2} = \frac{Th_2 \cdot B_2 \cdot Z_2}{\sin \beta_{B,2}}$$

where Z_2 might be different to Z_1 when splitter blades are used.

Meridional and tangential components of absolute velocity at the rotor trailing edge (Figure 4.3):

$$C_{M,2} = \frac{\dot{m}}{\rho_2 \cdot A_2} \tag{4.9}$$

Tangential component of relative velocity:

$$W_{U,2} = C_{M,2} \cdot \tan \beta_{B,2} + slip \tag{4.10}$$

$$U_2 = 2\pi R_{rms,2} \cdot N$$

$$slip = C_{U,2_{TH}} - C_{U,2} = U_2 \cdot (1 - \sigma)$$

where

$$\sigma = 1 - \frac{slip}{U_2}$$

$\sigma = \sigma_{ratio} \cdot \sigma_{design}$ is defined as the slip factor $\sigma_{ratio} = 1$ in the design condition, while σ_{design} is predicted using the Pfleiderer correlation (Pfleiderer 1952), calculated from the

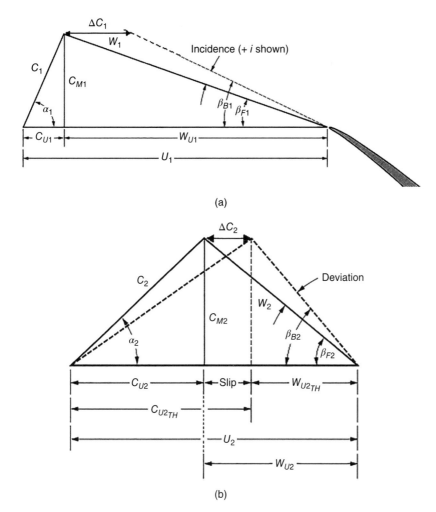

Figure 4.3 Rotor inlet (a) and exit (b) velocity diagram (not to scale). Source: Reproduced with permission from Veres (1994).

rotor geometry using the equation:

$$\frac{1}{\sigma_{design}} = 1 + \frac{1 + 0.6 \cdot \sin \beta_{B,2}}{(Z_2 \cdot (1 + \delta) \cdot X^2 + 0.25 \cdot (1 - \delta)^2)^{0.5}} \quad (4.11)$$

where $X = S/R_{rms}$, $\delta = R_{rms1}/R_{rms2}$, S is the blade length from inlet to exit midspan measured along the middle line of the blade (Figure 4.2), and R_{rms1} and R_{rms2} are the root-mean-square radii values calculated between the hub and tip, respectively. An additional correction factor is applied at off-design operation in order to account for changes in slip from its design value. This additional correction factor can be calculated using the following polynomial, obtained from Veres (1994) using a simple cubic regression:

$$\sigma_{ratio} = \frac{\sigma}{\sigma_{design}} = 1.534988 - 0.6681668 \cdot F + 0.077472 \cdot F^2 + 0.0571508 \cdot F^3 \quad (4.12)$$

Rotor exit relative flow angle and deviation, that is, difference between the flow relative angle and the blade angle at the rotor exit:

$$\beta_{F,2} = \arctan\left(\frac{C_{M,2}}{W_{U,2}}\right), \quad \text{deviation} = \beta_{B,2} - \beta_{F,2} \tag{4.13}$$

4.1.2.1.4 Impeller Off-Design Operation

The off-design flow-speed ratio is used to determine the equivalent flow coefficient in off-design conditions:

$$F = \frac{(Q/N)_{off-design}}{(Q/N)_{design}} \tag{4.14}$$

4.1.2.1.5 Suction Performance

In this model, cavitation onset is estimated based on assumptions on the blade-to-blade loading parameter BB and using the local value of static pressure at the pump leading edge:

$$P_{S,throat} = P_{T,1} - \frac{(C_1 \cdot BB)^2}{2}\rho \tag{4.15}$$

This value is compared to the local vapor pressure P_v to check for the onset of cavitation. Values for the loading parameter BB, which considers how much the flow is tangentially deviated by the pump impeller (i.e., how large the pressure drop inside the impeller is), can be in the range of 1.0–1.3 at design conditions. Onset of pump cavitation within 10% accuracy is believed to correspond to $BB = 1.25$.

Pump $NPSH$ and the pump's suction-specific speed:

$$NPSH = \frac{P_{T,1} - P_v}{\rho \cdot g}, \quad Nss = \frac{N \cdot Q^{0.5}}{NPSH^{0.75}} \tag{4.16}$$

Off-design suction performance is estimated using the suction-specific speed curve depicted in Figure 4.4, which is normalized and referenced to the design point suction capability.

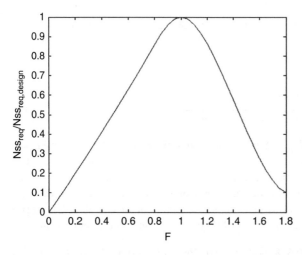

Figure 4.4 Suction performance at off-design flow-speed parameter. Source: Reproduced with permission from Veres (1994).

To avoid cavitation and pump stall the following conditions must be fulfilled:

$$Nss < \frac{Nss_{req}}{Nss_{req,design}} \cdot Nss_{req,design}, \; P_{S,throat} > P_v \qquad (4.17)$$

4.1.2.1.6 Impeller Performance

Eulerian head rise through the rotor is calculated as follows:

$$gH_{2,t} = U_2 \cdot C_{U,2} - U_1 \cdot C_{U,1} \qquad (4.18)$$

The impeller hydraulic efficiency is calculated as the ratio of actual head H_2 to the Eulerian head $H_{2,t}$,

$$H_2 = H_{2,t} \cdot \eta_{hyd} \qquad (4.19)$$

As suggested in Veres (1994), the impeller efficiency values are empirically derived based on correlations (see later on).

The pump's specific speed is:

$$k = \frac{2\pi N \cdot Q^{0.5}}{(g \cdot H_2)^{0.75}} \qquad (4.20)$$

The BEP for rotor hydraulic efficiency in terms of total-to-total conditions can be estimated from the following correlations:

$$\begin{cases} \eta_{hyd,design} = -0.6898 \cdot k^2 + 0.9837 \cdot k + 0.5785 \; for \; k \leq 0.8 \\ \eta_{hyd,design} = 1.02 - 0.12 \cdot k \; for \; k > 0.8 \end{cases} \qquad (4.21)$$

Off-design variation of impeller efficiency is empirically derived from pump data (Stepanoff 1948) for values of F between 0.7 and 1.2, while the limits of the curve are extrapolated from Veres (1994):

$$\eta_{hyd,ratio} = \frac{\eta_{hyd}}{\eta_{hyd,design}} = 0.86387 + 0.3096 \cdot F - 0.14086 \cdot F^2 - 0.029265 \cdot F^3 \qquad (4.22)$$

Thus $\eta_{hyd} = \eta_{hyd, design} \cdot \eta_{hyd, ratio}$

Substituting formulas Eq. (4.18), (4.19), (4.21), and (4.22) into Eq. (4.20), the implicit equation depending on k is obtained:

$$f(k) = \frac{2\pi N \cdot Q^{0.5}}{(U_2 \cdot C_{u,2} - U_1 \cdot C_{u,1})^{0.75} \cdot (\eta_{hyd,ratio} \cdot \eta_{hyd,design}(k))^{0.75}} - k = 0 \qquad (4.23)$$

The solution of Eq. (4.23) gives the pump dimensionless specific speed k, and then, using Eq. (4.21) and Eq. (4.22), the calculations of the ideal hydraulic efficiency of the pump and, ultimately, the impeller head rise are possible.

It is worth underlining that the validity of efficiency correlations is based on preliminary choice of pump configuration, but the generalization of such a correlation might not be correct. For this purpose, the following approach is used (Stepanoff 1948). Pump losses can be divided into concentrated h_c and distributed h_d:

$$H_2 = H_{2,t} - (h_c + h_d) \qquad (4.24)$$

where $h_c = h_{c, in} + h_{c, out}$ and $h_d = k_{d,rel} \cdot \frac{w_1^2}{2} + k_{d,fix} \cdot \frac{C_{M,1}^2}{2}$

Concentrated losses due to sudden deviation of flow at impeller inlet and exit are:

$$h_{c,in} = k_{c,in} \cdot \frac{\Delta C_1^2}{2}, h_{c,out} = k_{c,out} \cdot \frac{\Delta C_2^2}{2} \tag{4.25}$$

$$\Delta C_2 = \frac{(U_2 + W_{U,2}) \cdot \tan(\alpha_3) - C_{M,2}}{\tan(\alpha_3)}$$

Coefficients $k_{c,in}$ and $k_{c,out}$ are indeed subject to calibration, tentative ranges for them are 0.01–0.05 and 0.04–0.05, respectively. Similarly, coefficients $k_{d,rel}$, $k_{d,fix}$ must result from careful validation where some guess values are found in the ranges of 0.03–0.08 and 0.02–0.05, respectively.

In fact, concentrated losses due to sudden deviation of flow at impeller exit can be estimated as these occur around the diffuser blades or near the volute tongue with a metal angle α_3 from the tangential direction.

Thus, it is possible to determine the actual hydraulic efficiency.

$$\eta_{hyd} = \frac{H_{2,t} - (h_c + h_d)}{H_{2,t}} \tag{4.26}$$

4.1.2.1.7 Diffusion System Pressure Recovery and Loss

Flow velocity at the volute throat of the vaned diffuser:

$$C_{throat} = \frac{\dot{m}}{\rho \cdot A_{throat}} \tag{4.27}$$

Velocity at the vaneless diffuser exit:

$C_3 = \sqrt{C_{U,3}^2 + \left(\frac{\dot{m}}{\rho \cdot A_3}\right)^2}$, where, owing to the angular momentum equation $C_{U,3} = C_{U,2} R_2 / R_3$.

4.1.2.1.8 Diffusion Loading Parameter

$$L = \frac{C_{throat}}{C_3} \tag{4.28}$$

Total pressure loss coefficient of the diffusion system design point $\omega_{2-4, design}$ is assumed to be known:

$$\omega_{2-4} = \frac{P_{T,2} - P_{T,4}}{P_{T,2} - P_{S,2}} \tag{4.29}$$

If the pump impeller and the volute are adequately matched, the minimum (i.e., design) value of the loss coefficient of the diffusion system can be in the range of 0.15–0.25 for pumps equipped with vaneless diffusers and volutes.

Variation in total pressure loss coefficient with loading is given by the following polynomial, obtained by interpolation of the empirical data provided in Veres (1994):

$$\omega_{2-4,ratio} = \frac{\omega_{2-4}}{\omega_{2-4,design}} = 1.8151 - 1.83527 \cdot L + 0.8798 \cdot L^2 + 0.18765 \cdot L^3 \tag{4.30}$$

$$\omega_{2-4} = \omega_{2-4,ratio} \cdot \omega_{2-4,design} \tag{4.31}$$

4.1.2.1.9 Stage Head and Power

Head rise through the pump stage:

$$H_4 = \frac{(P_{T,4} - P_{T,1})}{\rho \cdot g} \tag{4.32}$$

Volumetric efficiency is based on internal (forward and backward) leakages and is expressed as the ratio of leakage to the inlet flow:

$$\eta_{vol} = \frac{Q}{Q + Q_f} \tag{4.33}$$

where $Q_f = Q_{f,for} + Q_{f,back}$

$$Q_{f,for} = 0.22\pi(2R_{f,for})S_{f,for}\sqrt{\frac{4gH_2}{1.5 + K_f\frac{B_{f,for}}{2S_{f,for}}}} \tag{4.34}$$

$$Q_{f,back} = 0.22\pi(2R_{f,back})S_{f,back}\sqrt{\frac{4gH_2}{1.5 + K_f\frac{B_{f,back}}{2S_{f,back}}}} \tag{4.35}$$

where K_f is a friction coefficient in the order of 0.1.

Mechanical efficiency (correlations are taken from Stepanoff 1948):

$$\eta_{mech} = \frac{Pow_{fluid}}{Pow_{fluid} + Pow_{mech}}, \text{ where } Pow_{fluid} = \frac{\dot{m} \cdot H_2 \cdot g}{\eta_{hyd}} \tag{4.36}$$

Disk friction loss is calculated using the following equation by using an experimental-based loss factor, K (Stepanoff 1948).

$$Pow_d = K \cdot N^3 \cdot R^5_{hub,2}, \text{ where } K = 3.953 \cdot 10^{-6} \tag{4.37}$$

Stage power required to drive the pump:

$$Pow_{stage} = \frac{\dot{m} \cdot H_2 \cdot g}{\eta_{hyd} \cdot \eta_{mech} \cdot \eta_{vol}} + Pow_d \tag{4.38}$$

A calibrated pump model can significantly improve the accuracy of predicted pump performance. However, the development of a common set of calibration parameters is likely to be obtained in order to achieve the best model reliability and accuracy. A simple way to do this consists of taking the mean value of the calibrated parameters.

A check of model accuracy after calibration is performed and the results obtained on sample types of radial pumps are given in Figure 4.5. As is apparent, the results obtained using the global pump model are good for preliminary design purposes.

Using a greater number of real pump data (a wider sample), it should be possible to find more accurate correlations between the calibration parameters and one or more indexes, based directly on the pump specifications instead of calculating the mean value of each parameter.

4.1.2.1.10 Considerations About 1D Models

A simple meanline model cannot simulate a real pump closely by matching the real pump's behavior. However, it is possible to calibrate such a model using real pump data by trying to reduce the performance prediction errors as much as possible. Test data should be superimposed onto the model results applied on the same pumps.

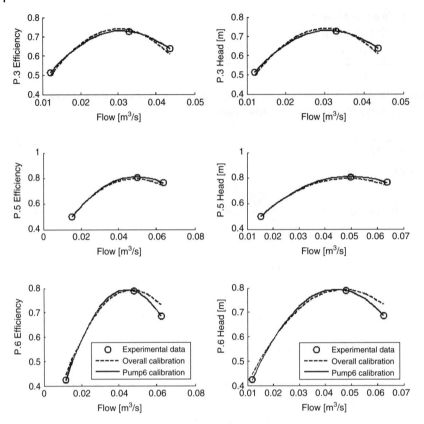

Figure 4.5 Calibrated pump trends represented by continuous lines are compared with the same trends, obtained by using the mean value of all the calibration parameters of the turbopumps 3-5-6, represented by dashed lines. Source: Reproduced with permission from Benini and Cenzon (2009), (figure 6 from original source), Copyright © 2009 by Institution of Mechanical Engineers.

A calibration procedure can be carried out using several methods. One of these is presented in Benini and Cenzon (2009), which consists of an optimization procedure that searches for the best matching between numerical and experimental data available using genetic algorithms (GA).

In the model calibration, the set of decision variables are in fact the coefficients of the loss correlations used through the meanline model at on/off-design conditions, and can be considered as genes forming a genome. For each pump, different genomes (genotype) produce different pump performance maps (phenotype). The best phenotype is the one that is closer to the real experimental data, thus the goal is to search for the best set of parameters that produces the minimum distance in the performance map from the experimental data. This is a typical optimization problem that can be solved using the previously described GA, where the genomes are the unknown calibration coefficients, each of which is defined within a prescribed range. The reader is referred to Benini and Cenzon (2009) for a detailed description of the method.

4.1.2.2 2D Models

Two-dimensional models are used in the intermediate design phases of a centrifugal pump design when more detailed evaluations are needed to assess pump performance.

In axial-flow pumps, 2D models are often necessary as on/off performance is more closely related to airfoil hydrodynamic characteristics compared to centrifugal types. Moreover, when multistage machines are accounted for, particularly of the axial type, 2D methods are still preferable to purely 3D ones as the latter are very expensive to run. Finally, 2D models are still very useful in multi-point optimization problems when the design focus is given on cascade angle.

The starting point for 2D analyses relies on the decomposition of the actual flow field surfaces into relative stream surfaces according to the fundamental quasi-3D theory by Wu (1952). Such stream surfaces are universally known under the name of S1 and S2 surfaces (see Figure 4.6).

As is well-known, obtaining the actual shape and pertinent flow field in either S1 or S2 surfaces is far from being easy as this requires an iterative procedure, mostly supported by computers. This is caused either by the fact that a numerical solution of systems of partial differential equations is needed on both surfaces and the fact that the flow on a surface family is correlated to the one occurring on the other family.

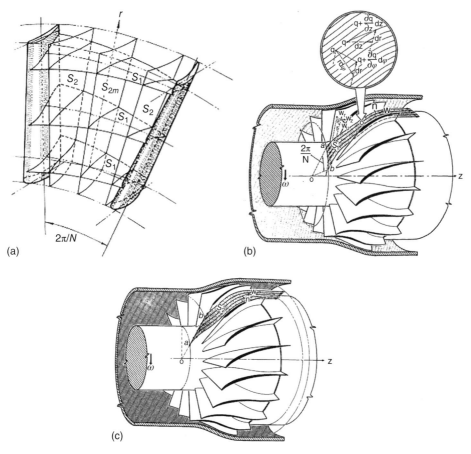

Figure 4.6 (a) S1 and S2 stream surfaces according to Wu for (b) an axial and (c) mixed-flow turbomachine. Source: Reproduced with permission from Wu (1952).

In a quasi-3D optimization framework, the flow solution over either S1 or S2 is searched for. Today, with the increasing availability of large and fast computers, a quasi-3D solution is, however, no longer so attractive for all types of turbomachines. In most practical cases of centrifugal machines, in fact, accurate 3D solutions of the flow field are relatively cheap to obtain nowadays and many engineers consider a quasi-3D method not worth considering as it is not cost-effective compared to pure 3D computations. This also holds true for compressible flow machines.

On the other hand, considerable attention is given to quasi-3D methods in axial machines, especially when they make use of many stages. These methods are usually classified into two categories: (i) throughflow and (ii) blade-to-blade methods; such methods can be valuably used both for incompressible and compressible flow turbomachines.

4.1.2.2.1 Throughflow Methods

Throughflow calculations make use of the Euler equations of motion (i.e., continuity, energy, inviscid momentum equation, and an equation of state – when compressible flows are dealt with) applied to the mean axisymmetric stream surface, which in fact corresponds to the S_{2m} surface in Figure 4.6. S_{2m} surfaces are the ones located about midway between two consecutive blades dividing the mass flow in the channel into two approximately equal parts (Novak and Hearsey 1977). For pump blades with radial elements, as long as the twist of the surface is not large, it is convenient to consider a mean stream surface formed by fluid particles originally lying on a radial line *ab* upstream of the blade row (see Figure 4.6), otherwise the radial line is chosen about midway in the passage with the fluid particles originally starting out from a curved line upstream of the blade row.

In throughflow calculations, effects of flow viscosity and related losses are usually included by using empirical modifications for gas entropy in the equation of state (usually in the form of loss, deviation, and blockage correlations).

Throughflow solutions can be obtained numerically by solving the previously mentioned equations using three methods: (i) the streamline curvature (SLC) method, (ii) matrix-based throughflow method, and (iii) time-marching methods. Among these, the SLC method was the first to be developed and is by far the most used, at least for low-speed and incompressible-flow turbomachines.

SLC methods require a grid in which the meridional channel of a turbomachine is discretized (Figure 4.7). Such a grid is based on the intersections (nodes) between two families of curves: (i) the meridional streamlines of the circumferentially averaged flow, and (ii) a priori defined stations that are approximately orthogonal to the streamlines (often referred to as quasi-orthogonals). The streamlined shapes of type (i) are dynamic over iterations (exception given for streamlines onto the hub and the shroud), as they change once a final converged solution is found. On the other hand, the calculating stations of type (ii) are somehow oriented so as to mimic the blade's leading and trailing edges everywhere in the flow field. The computational results are obtained iteratively on each node where a numerical solution for the radial equilibrium (RE) equation is searched for.

The reader is referred to Denton (1978) for a rather complete review of throughflow methods, as these are outside the scope of this book. A useful step-by-step procedure for the SLC method over a generic turbomachine is given in Schobeiri (2005).

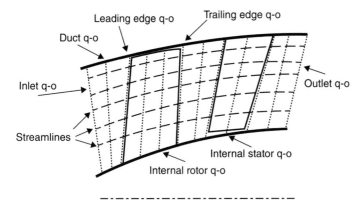

Figure 4.7 Computational grid for throughflow calculations in a turbomachinery stage.

Using throughflow methods, it is possible to manage several decision variables, including: hub and shroud contours (including their radii at each quasi-orthogonal stations), flow angles, and velocities along the span (from which blade angles can be obtained based on deviation correlations). If proper loss correlations are also implemented, such methods can give the distribution of efficiency, as well as total enthalpy, either in absolute or relative frames of reference in the spanwise direction. Therefore, throughflow methods are well suited to optimization purposes as they can properly deal with shape variables influencing machine performance, including characteristic maps.

4.1.2.2.2 Blade-to-Blade Methods

Blade-to-blade methods may be applied either as simple approximations or more accurate solutions of the blade-to-blade flow equations. Accurate methods are applicable when the exact locations of the S1 surfaces of the flow are known. Since this involves the obtainment of a converged coupled S1 + S2 flow solution, accurate methods are of limited interest today as the effort required to get a fully converged coupled S1 + S2 simulation is comparable to that needed for a CFD 3D solution, which in general provides more accurate results.

Quasi-3D blade-to-blade methods define the cascade geometry on an axis-symmetric surface (surface of revolution R about the z-axis in Figure 4.8).

The link between the conventional cylindrical frame of reference (r, θ, z) and conformal coordinates (m', θ, z) is given by the following equations:

$$
\begin{cases}
m' = \int \dfrac{dm}{r} = \int \dfrac{\sqrt{dr^2 + dz^2}}{r} \\[2ex]
\theta = \theta \\[2ex]
z = \int \dfrac{dz}{dm'} dm' = \int \pm\sqrt{r^2 - \left(\dfrac{dr}{dm'}\right)^2} \, dm'
\end{cases}
\tag{4.39}
$$

From which it is possible to calculate the local blade angle: $\tan \beta = (r d\theta)/m = d\theta/m'$ and based on the curve length increment ds in the R surface $ds = \sqrt{dr^2 + dz^2 + (r\,d\theta)^2}$ $= \sqrt{dm^2 + (r\,d\theta)^2} = r\sqrt{dm'^2 + d\theta^2}$, the total length S of any curve on the R surface

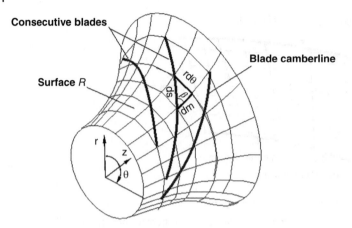

Figure 4.8 Conformal mapping.

(e.g., the total length of a camberline that is curved in space, see Figure 4.8):

$$S = \int ds = \int r\sqrt{dm'^2 + d\theta^2} \tag{4.40}$$

Finally, the mapping from cylindrical to Cartesian coordinates can be easily accomplished when necessary:

$$x = \pm \frac{r}{\sqrt{1 + \tan^2\theta}}$$

$$y = x \tan \theta$$

$$\theta = \theta \tag{4.41}$$

On the other hand, simple 2D blade-to-blade methods generally use a strong (yet consistent for most axial turbomachines) approximation regarding the RE equation outside the blade rows, that is, that the radial component of flow velocity is negligible. Such a hypothesis leads to the simplified RE equation:

$$\frac{1}{\rho}\frac{dh^0}{dr} = c_x\frac{dc_x}{dr} + \frac{c_\theta}{r}\frac{d(rc_\theta)}{dr} \tag{4.42}$$

From which, when a proper (or tentative) swirl velocity distribution along the radius is given, the spanwise distribution of velocity triangles can be obtained. This approximation leads to optimization methods that are, strictly speaking, 2D in nature. The flow computation over the blade-to-blade plane of single or multiple blade rows can be performed using CFD codes.

An example of application of such methods in the blade-to-blade design of axial-flow pump cascades is given in Benini and Toffolo (2001). These methods require the definition of a 2D flow domain (see Figure 4.9) for where to specify the decision variables of the optimization. These may include basically all cascade parameters; that is, airfoil shape, chord, spacing (or the solidity), and stagger angle. Blade spacing is often determined by number of blades, which is established in advance, while the chord is somewhat determined by practical considerations on blade solidity.

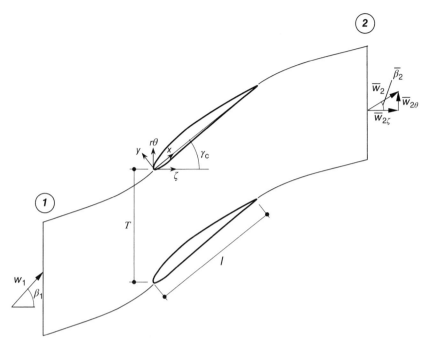

Figure 4.9 A 2D blade to blade model of an axil flow pump cascade. Source: Reproduced with permission from Benini and Toffolo (2001) (figure 5 from original source), Copyright © 2001 by Institution of Mechanical Engineers.

Blade shape is by no means parameterized using either explicit or parametric curves. Explicit functions are of the type that follows (NACA 4-digit airfoils):

$$
\begin{aligned}
y_c &= \frac{m}{p^2}(2px - x^2) && \text{from } x = 0 \text{ to } x = p \\
y_c &= \frac{m}{(1-p)^2}[(1 - 2p) + 2px - x^2] && \text{from } x = p \text{ to } x = c
\end{aligned}
\tag{4.43}
$$

where maximum camber (m) is given as a percentage of the chord length (along which the independent variable x goes from 0 to c), the maximum camber (p) is expressed in tenths of the chord length for the mean camberline, and t is the maximum thickness that defines the thickness distribution above ($+$) and below ($-$):

$$
\pm\, y_t = \frac{t}{0.2}\left(0.2969\sqrt{x} - 0.1260\,x - 0.3516\,x^2 + 0.2843\,x^3 - 0.1015\,x^4\right)
\tag{4.44}
$$

These expressions can be generalized using almost arbitrarily defined suction and pressure-side polynomials; for example, as follows:

$$
\begin{aligned}
\pm\, y_t &= a_0\sqrt{x} + a_1 x + a_2 x^2 + a_3 x^3 && \text{ahead of } t_{max} \\
\pm\, y_t &= d_0 + d_1(1 - x) + d_2(1 - x)^2 + d_3(1 - x)^3 && \text{aft of } t_{max}
\end{aligned}
\tag{4.45}
$$

where the polynomial coefficients $\mathbf{a_i}$-s and $\mathbf{d_i}$-s can become the decision variables of the optimization.

Parametric curves are today extensively used for airfoil parametrization. The most common way to do this is perhaps to use n-order Bézier curves:

$$
\begin{cases}
x(t) = \sum_{i=0}^{n} C_n^i t^i (1-t)^{n-i} x_i \\
y(t) = \sum_{i=0}^{n} C_n^i t^i (1-t)^{n-i} y_i
\end{cases}
\tag{4.46}
$$

where t identifies the curve parameter varying in $[0,1]$, x_i and y_i are the normalized coordinates of the control points, and $C_n^i = n! / (i!(n-i)!)$. One curve is used for the suction side and another for the pressure side. As documented in Benini and Toffolo (2001), the abscissas x_i of the control points are often uniformly spaced in $[0,1]$, while the ordinates y_i are chosen inside a prescribed range of possible values. A set of 10 parameters is selected for the complete parameterization of the airfoil shape; then two further parameters are needed to define the length of the chord (with respect to which the coordinates are normalized) and the stagger angle, if necessary.

4.1.2.3 3D Models

3D fluid dynamic models are often used in the final stages of pump design as the computational effort required to run iterative 3D analyses on a wide search space is not compatible with the development duration at an industrial level. Optimization is almost invariably restricted to BEP conditions, while multi-point optimization is generally avoided and assessment of off-design behavior treated as a post check.

In 3D models, it is common to start with a baseline solution whose geometry has been previously modeled using CAD software (Figure 4.10). Impeller geometry is usually described using a combination of parametric S1 and S2 surfaces: parametric curves are used almost independently on approximated S1 and S2 surfaces for a desired number of blade sections (layers) using curves that have been already presented in the previous paragraphs. Then, the 3D blade geometry is reconstructed using loft (or blend) operations in the CAD software, which interpolate the profiles coordinates along the spanwise direction defined as the path for lofting.

Stationary components such as volutes are easily parameterized using algebraic equations that define the radial coordinate r_A as a function of the ε angle (Figure 4.11), as well as with simple equations for $b = f(r)$. A typical law for r_A is the following:

$$
r_A = A e^{B\varepsilon}
\tag{4.47}
$$

where the coefficients A and B are taken as decision variables. Additional decision variables are r_Z, the value of b and the angle α_{3B} in Figure 4.11. In particular cases (Figure 4.11), a further design variable δ is useful to help defining the shape of volute cross section.

Following this, a CFD model is set up with all the pertinent boundary conditions, including both static and rotating (or dynamic domains). The number of blade channels to be simulated depends on the type of rotor-stator interaction interface scheme. To this purpose, usually three approaches are commonly available in CFD codes: (i) mixing plane, (ii) MRF (Multiple Reference Frame) or frozen rotor, and (iii) sliding meshes (Vande Voorde et al. 2004). The first two are steady in nature, while the third

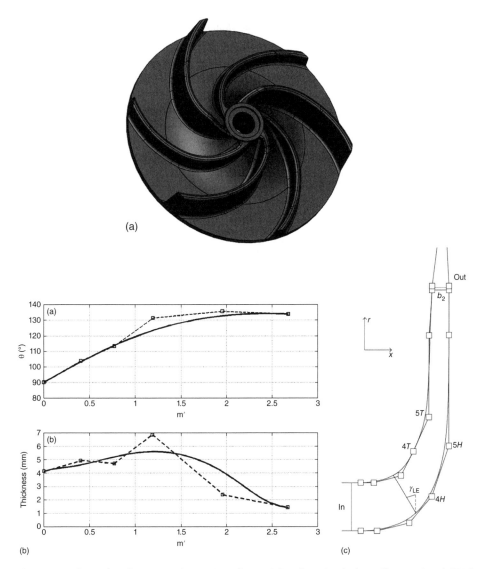

Figure 4.10 Examples of parameterization impeller meridional section (a, from Checcucci et al. 2011), of blade camber line (b) and thickness distribution using Bézier polynomials (c).

involves unsteady calculations. Examples of typical boundary conditions applied to either a single-channel impeller or a complete impeller-volute configuration is given in Figure 4.12.

In optimization problems, either the mixing plane or MRF approach are preferred to unsteady sliding meshes in view of their far lesser computational effort. In MRF methods, a fixed position is given to the rotor system; therefore, the equations for rotor flow motion are solved in a rotating reference frame, where both centrifugal and Coriolis forces are accounted for. Regarding the stator, governing equations are solved in an absolute reference frame.

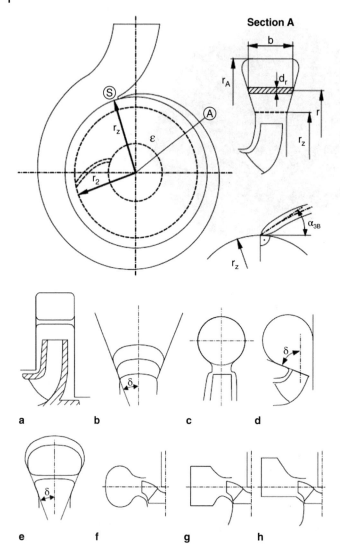

Figure 4.11 Pump volute parameterization.

Mesh sensitivities studies are always fundamental to establish the best trade-off between model accuracy and CPU time. Structured, unstructured, and mixed meshes can be profitably used, although mixed-types (i.e., hexahedral elements near walls and tetrahedral elements in the core flow regions) provide more flexibility in iterative loops while keeping a very good level of accuracy.

Regarding CFD codes, RANS solvers are commonly used, while large eddy simulation (LES) or even more sophisticated approaches using alternative filtering techniques for the Navier–Stokes equations are today impractical for solving optimization problems at an industrial level. As far as turbulence models are concerned, successful calculations are documented using the standard k-ε or k-ω models, although the k-ω shear stress transport (SST) turbulence model usually provides better results compared to the standard or

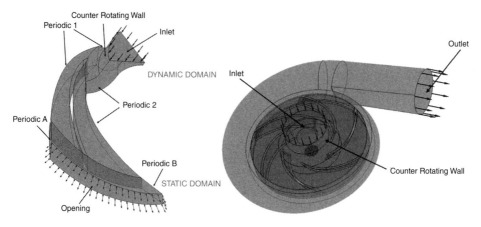

Figure 4.12 Example of computational domain of a centrifugal pump with volute.

RNG model. In principle, cavitation models can also be activated, where homogeneous two-phase RANS equations can be used. Once again, to the authors' best knowledge these approaches, although available nowadays, are not of practical utilization in complex optimization loops.

Due to the significant computational effort, 3D optimization is often carried out using surrogate-assisted algorithms. Typical procedures involve:

- Initial exploration of the search space using direct Design of Experiment (DoE) methods (global search);
- Surrogate, or metamodel, construction;
- Sensitivity or significance analysis;
- Search for approximate optima on the surrogate (local search);
- (Optional) local refinement of surrogate model using infill criteria around approximate optima;
- (Optional) high-fidelity optimization on surrogate.

DoE term refers to an exploratory technique based on statistical concepts to identify the impact that contemporary variation of different factors has on the final result (Anderson 1997). In the context of optimization, the purpose of DoE is to identify the most effective numerical experiments that lead to the best results in terms of fitness function knowledge using the minimum number of samples to be simulated. Sampling is therefore essential in determining DoE effectiveness. Two techniques for sampling are mainly used today: the Monte Carlo Sampling method (Robert and Casella 2005) and Latin Hypercube Sampling (LHS) (McKay et al. 1979). The second is an evolution of the first that requires a high number of samples to effectively approximate the actual distribution of variables. Both techniques are based on the cumulative distribution function $P(x)$:

$$P(x) = \int_{-\infty}^{x} p(x)dx \tag{4.48}$$

$$p(x) = \frac{1}{\sigma\sqrt{2\pi}} \exp\left[-\frac{(x-\mu)^2}{2\sigma^2}\right]$$

for a Gaussian distribution with a μ mean, a σ standard deviation of the distribution, and x any measured value.

In the Monte Carlo method, sampling occurs randomly from the probability distribution, which indeed may lead to a clustering of samples around the mean value of a variable (e.g., in Figure 4.13 sampling would lead to a bias toward values in the neighborhood of $x = 0$ since there the cumulative probability distribution has the steepest slope). This is regarded as non-optimal as it discards the region of lower cumulative probability.

In the LHS method, sampling is carried out by subdividing the cumulative probability range [0, 1] into intervals having equal partitions' amplitude (Figure 4.14). This is often referred to as a "stratified sampling." LHS mitigates the clustering of samples as these are more uniformly spread throughout the search space of a single variable.

The extension of both Monte Carlo and LHS methods to multivariate problems is straightforward. Variants of LHS (such as Optimal Space Filling Design) involve maximization of the reciprocal distance between samples in the decision variable space and this assures an even more apparent effectiveness of DoE (Figure 4.15).

4.1.2.3.1 Sensitivity Analyses

Sensitivity analyses are of primary importance to understand the significance of decision variables with respect to objective functions. Such analyses can be carried out to establish a ranking of the parameters in order of influence and can be done either prior or just after the construction of metamodel. To this purpose, global sensitivity techniques are often used (Saltelli et al. 2008), such as correlation analysis.

4.1.2.3.2 Univariate Analysis

Correlation analysis computes the correlation coefficient $R \in [-1, 1]$, which is a quantitative indicator of how a model parameter value x_i and the pertinent objective function

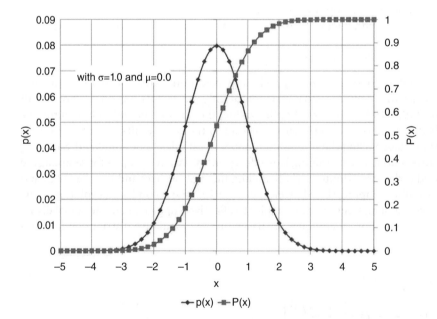

Figure 4.13 Probability distribution function [p(x)] and cumulative probability distribution function [P(x)] for a Gaussian distribution (for a 0 mean and unitary standard deviation).

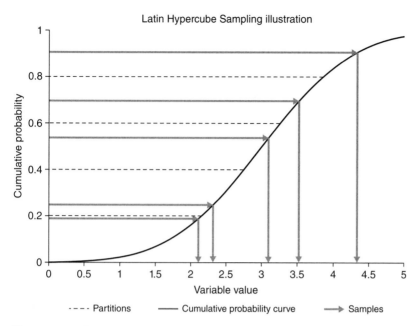

Figure 4.14 LHS criterion.

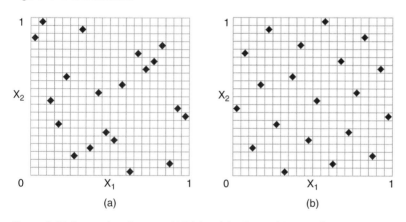

Figure 4.15 Comparison between LHS (a) and the Optimal Space Filling Design (b).

outputs y_i are correlated, being $i = 1,...,n$. A popular indicator is known as Pearson's correlation coefficient:

$$R(X, Y) = \frac{\sigma_{XY}}{\sigma(X)\, \sigma(Y)} \tag{4.49}$$

where

$$\sigma_{XY} = \text{cov}(X, Y) = \frac{1}{n-1} \sum_{i=1}^{n} (x_i - \mathbb{E}(X))(y_i - \mathbb{E}(Y))$$

$$= \frac{1}{n-1} \sum_{i=1}^{n} (x_i - \mu(X))(y_i - \mu(Y)) \tag{4.50}$$

and

- "cov" is the covariance function, $\mathbb{E}$ is the expected value, or the mean μ in the case of normally distributed probability variables.
- $\sigma(X)$ and $\sigma(Y)$ are the standard deviations for X and Y, respectively:

$$\sigma(X) = \sqrt{\frac{1}{n-1} \sum_{i=1}^{n} (x_i - \mu(X))^2}, \sigma(Y) = \sqrt{\frac{1}{n-1} \sum_{i=1}^{n} (y_i - \mu(Y))^2}.$$

Negative values of R mean negative (or inverse) correlations, while positive values indicate positive correlations. Finally, when R is 0, or very close to 0, the variable and the objective functions are said to be uncorrelated. The highest absolute values of R indicate the strongest correlations.

An example of results obtained from correlation analysis is depicted in Figure 4.16. From an almost uncorrelated data pairs (X is the decision variable and Y the associated fitness function) on the left-hand side of the figure, a weak inverse correlation is found in the middle plot and a very apparent positive correlation is registered on the right-hand side of the figure.

4.1.2.3.3 Multivariate Analysis

Correlation analysis is useful also to discover mutual correlation among the decision variables. If $\mathbf{X}$ is the vector of decision variables linked to a given value of the fitness function, then we can calculate the expected value for the i-th decision variable simulated m times:

$$\mu_i = \mathbb{E}(X_i) = \frac{1}{m} \sum_{k=1}^{m} X_{ik} \tag{4.51}$$

and, element Σ_{ij} of the variance-covariance matrix Σ becomes:

$$\Sigma_{ij} = \text{cov}(X_i, X_j) = \mathbb{E}[(X_i - \mu_i)(X_j - \mu_j)] \tag{4.52}$$

In other words,

$$\Sigma = \begin{bmatrix} \mathbb{E}[(X_1 - \mu_1)(X_1 - \mu_1)] & \mathbb{E}[(X_1 - \mu_1)(X_2 - \mu_2)] & \cdots & \mathbb{E}[(X_1 - \mu_1)(X_n - \mu_n)] \\ \mathbb{E}[(X_2 - \mu_2)(X_1 - \mu_1)] & \mathbb{E}[(X_2 - \mu_2)(X_2 - \mu_2)] & \cdots & \mathbb{E}[(X_2 - \mu_2)(X_n - \mu_n)] \\ \vdots & \vdots & \ddots & \vdots \\ \mathbb{E}[(X_n - \mu_n)(X_1 - \mu_1)] & \mathbb{E}[(X_n - \mu_n)(X_2 - \mu_2)] & \cdots & \mathbb{E}[(X_n - \mu_n)(X_n - \mu_n)] \end{bmatrix} \tag{4.53}$$

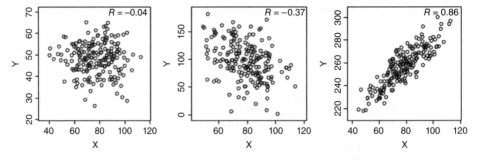

Figure 4.16 Examples of different correlations between two variables, X and Y.

From which a correlation matrix can be derived:

$$C = \text{corr}(X) = (\text{diag}(\Sigma))^{-\frac{1}{2}} \Sigma (\text{diag}(\Sigma))^{-\frac{1}{2}} \qquad (4.54)$$

where $\text{diag}(\Sigma)$ is the matrix of the diagonal elements of Σ. Each element on the principal diagonal of Σ gives the correlation coefficient of a decision variable with itself, which by definition equals 1. Each off-diagonal element of Σ gives the correlation coefficient of a decision variable with a different one and eventually quantifies mutual dependencies among variables as such an element that takes values in the range $[-1,1]$. The meaning of the value is analogous to the one seen for scalar decision variables. An example of a correlation matrix, in which each element of the matrix is shaded based on the correlation coefficient of a 34-decision variable problem, is given in Figure 4.17.

Another way to visualize a correlation matrix is to use a mixed graphical-quantitative plot, as depicted in Figure 4.18 for a sample six-decision variable problem. Since the correlation matrix is symmetrical, in the upper right part of the matrix the generic S_{ij} element contains the values of the correlation coefficient between variables X_i and X_j. In the corresponding S_{ji} element of the matrix, a graphical representation of the dependency between X_i and X_j is given.

4.1.2.3.4 Multivariate Multi-Objective Problems

Statistical tools, in particular correlation matrices, are useful to assess correlation among variables and decision variables in multivariate and multi-objective problems. By extending the concept of correlation matrix to include fitness functions in the statistical analysis, a powerful assessment of correlations is built that is known as the correlogram plot. An example of such plot is given in Figure 4.19, where a three-variable (VAR1, VAR2, and α), three-objective (CL, CD, and CM) problem is dealt with. For

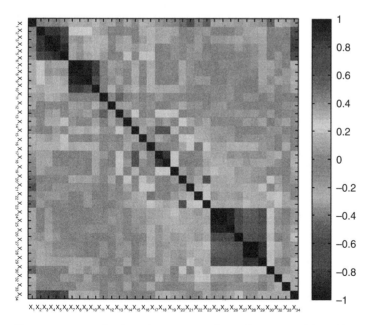

Figure 4.17 Example of correlation matrix of a 34-decision variable problem. Mutual correlation of decision variables is apparent in the different shaded regions.

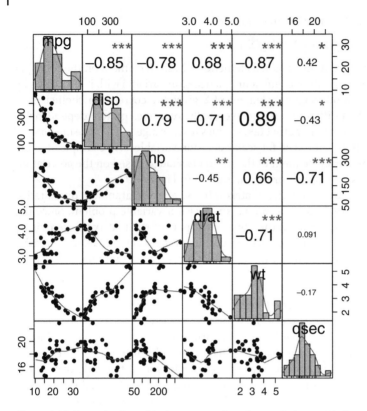

Figure 4.18 Example of graphical representation of a correlation matrix for a six-decision variable problem. Relevant mutual correlations of decision variables are apparent in cells with three gray stars.

the sake of clarity, it is worth mentioning that such example refers to an aerodynamic optimization of lift, drag, and moment coefficients of a streamlined body, whose shape is described using two geometric variables (VAR1 and VAR2) together with an operating variable, the angle of attack α. First of all, it can be observed that the two independent variables describing the shape, VAR1 and VAR2, are correlated to each other with a roughly linear relationship. However, this does not mean that VAR1 and VAR2 are non-significant but that they are mutually linked. This is typical when dealing with parameterization of complex geometries. Instead VAR1 and VAR2 are completely uncorrelated with the angle of attack as the related scatterplots are sparse and the pie charts empty.

Furthermore, the correlogram plot is helpful to assess influence of decision variables on objective functions. As far as the effect of the independent variables on the aerodynamic coefficients is concerned, both the lift, and the pitching moment are highly dependent on the incidence angle, as indicated by the filled portion of the corresponding pies on the right of the diagram; moreover, the relationship is of linear type, as illustrated by the scatter plots on the left of Figure 4.19, which are very close to the diagonal. There is clear evidence of a noticeable degree of correlation of the drag coefficient CD with α, even though less pronounced than the one of lift and moment: actually, a parabolic dependence of CD on α is recognizable in the scatterplot, even though a certain degree of

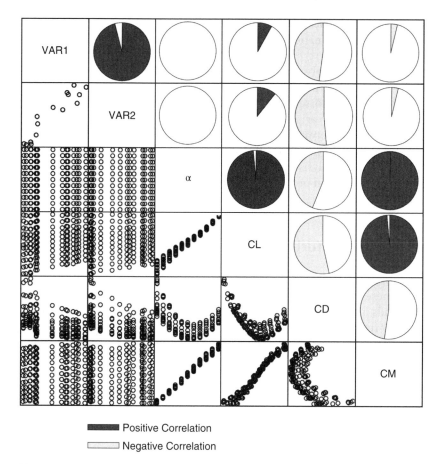

Figure 4.19 Correlogram plot including decision variables (VAR1, VAR2, α) and objective functions (CL, CD, CM).

sparseness in the point cloud is evidenced. Finally, regarding the impact of the geometric variables, a remarkable effect of the geometric shape is shown on the drag coefficient. In fact, a specific trend of the CD values with both VAR1 and VAR2 may be observed in the related scatterplots (with decreasing drag at increasing VAR1 and VAR2). On the other hand, VAR1 andVAR2 seem to have a low impact on determination of the lift coefficient and a negligible effect on the moment.

In summary, correlation analysis is useful to understand:

- The degree of correlation of a decision variable to the fitness function, that is to test hypotheses about cause-and-effect relationships between the two quantities. Associated with this, the value of one decision variable can be predicted in association with some properties of the fitness function, for example, its local optimality, based on a high absolute value of the correlation coefficient. Care must be taken in this respect, as the correlation coefficients are susceptible to sample selection biases that may lead to erroneous predictions. Finally, it is worth underlying that small correlation values do

not necessarily mean that a decision variable and its fitness function are poorly correlated. To get complete information, scatterplots should always be examined together
with correlation coefficients.

- The degree of correlation between pairs of decision variables does not necessarily
infer a cause-and-effect relationship. This property is of particular relevance in optimization problems, as the correlation methods presented here should be used only
with independent decision variables. In this respect, it is worth mentioning that a
high degree of correlation between variables might not indicate that they are mutually
influenced but that they are possibly influenced by one or more additional variables.
Therefore, a correlation analysis is always useful; for example, beforehand, a proper
optimization is carried out in order to highlight possible dependencies among decision variables.

4.2 Compressors and Turbines

Aerodynamic optimization techniques for compressors and turbines have gradually
changed the way these machines are today conceived and designed. While conventional
1D and 2D methods are well established for preliminary design calculations and are
being continuously refined for better reliability and fast search space explorations, 3D
investigations have been developed and are well-known within industries and academia
for more detailed designs.

Compressor and turbine design still remain very complex and multidisciplinary tasks
in which aero-thermodynamic drivers are more and more integrated with mechanical, technological, structural, and noise-related concerns. Such complexity requires a
well-organized and multi-criteria design methodology.

4.2.1 Axial, Radial, Multistage Compressors

Axial compressors are inherently multistage turbomachines, which are widely used to
compress large mass flow rates of compressible fluids with a relatively low-pressure
jump per single stage. Because of their multi-row configuration and consequent higher
costs compared to other compressor types, their utilization is almost universally confined to (i) large-volume air separation and treatment plants in industrial plants and (ii)
in both stationary and vehicular gas turbines. Axial compressors are in fact a unique
choice today in large gas turbines such as jet engines, high-speed ship propulsors, and
medium-to-large scale power stations.

Centrifugal compressors, often referred to as "radial" or "turbo" compressors, are one
of the most used turbomachines for rising the working pressure of compressible fluids
in industrial plants. Common applications involve natural gas compression in pipelines,
air compression in pneumatic circuits, air turbocharging and supercharging in automotive engines, and air compression in small-to-medium gas turbines. Similar to their
incompressible-flow counterpart, that is, centrifugal pumps, centrifugal compressors
are used when high-pressure rise is required compared to the mass flow rate being man-
aged by the component. Multistage configurations are also well-known and are used
when very high-pressure ratios are needed, such as in oil refineries, natural gas processing, petrochemical, and chemical plants. Because of the large power involved in such

utilization, inter-cooling is often employed between stages to control air temperature and reduce overall work absorption.

4.2.2 Parametric Shape Models and Flow Solvers for Axial Compressor Optimization

4.2.2.1 1D Models

A preliminary 1D design is usually performed on a single "design block" (Figure 4.20), to which basic thermo- and fluid dynamics equations are applied to each stage at the mean compressor diameter in which semi-empirical correlations are used for performance prediction. Next, compressor stages are stacked serially to establish the overall compressor performance neglecting stage interaction. Next to the 1D flowpath definition at the mean diameter, the designer can establish the distribution of stage parameters along the rotor and stator radii. The reader is referred to the work by Benini (2013) for a detailed description of 1D methods for compressor design.

An example of prediction using 1D meanline compressor model is given in Figure 4.21, where the map of the Rolls-Royce HP9 compressor stage (Ginder and Harris 1989) has been calculated by the authors and compared to published experimental data.

Despite its relative simplicity, meanline 1D methods based on stage-stacking techniques still play an important role in the design of compressor stages, as also demonstrated by Sun and Elder (1998). In their work, a numerical methodology is used for optimizing a stator stagger setting in a multistage axial-flow compressor environment (seven-stage aircraft compressor) based on a stage-by-stage model to "stack" the stages together with a dynamic surge prediction model. A direct search method incorporating a sequential weight increasing factor technique (SWIFT) was then used to optimize stagger setting, while the objective function was penalized externally with an updated factor that helped to accelerate convergence.

An example of how 1D models can still be used in the preliminary design optimization of axial compressors is given by Chen et al. (2005). In their work, a model for the optimum design of a compressor stage, assuming a fixed distribution of axial velocities,

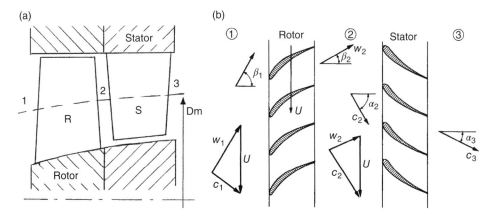

Figure 4.20 Sketch of a compressor stage (a) and cascade geometries at midspan (b).

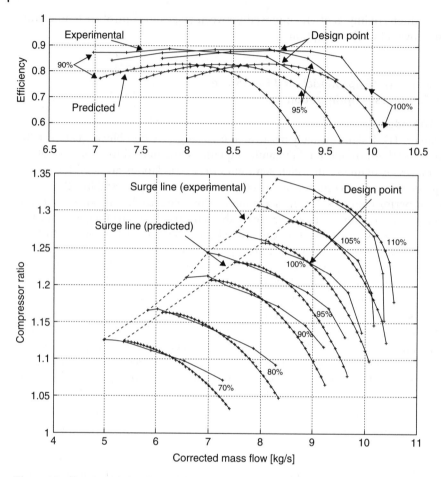

Figure 4.21 Experimental and predicted map of the Rolls-Royce HP-9 compressor.

is presented. The absolute inlet and exit angles of the rotor are taken as design variables. Analytical relations between the isentropic efficiency and the flow coefficient, the work coefficient, the flow angles, and the degree of reaction of the compressor stage were obtained. Numerical examples were provided to illustrate the effects of various parameters on the optimal performance of the compressor stage.

4.2.2.2 2D Models

A second preliminary step, distinct from the 1D procedure, is 2D design, which includes both cascade and throughflow models, from which a characterization of both design and off-design multistage compressor performance can be carried out after some iterations, if necessary. In this case, both direct and inverse design methodologies have been successfully applied. Numerical optimization strategies may be of great help in this case as the models involved are relatively simple to run on a computer. Often an optimization involves coupling with a prediction tool, for example, a blade-to-blade solver and/or a throughflow code.

4.2.2.3 Advanced Throughflow Design Techniques (2D)

Throughflow design allows configuring the meridional contours of the compressor, as well as all other stage properties in a more accurate way compared to 1D methods. They make use of cascade correlations for total pressure loss/flow deviation and are based on throughflow codes, which are two-dimensional inviscid methods that solve for axisymmetric flow (RE equations) in the axial-radial meridional plane (Figure 4.22). A distributed blade force is imposed to produce the desired flow turning, while the blockage factor accounts for the reduced area due to blade thickness, and distributed frictional force representing the entropy increase due to viscous stresses and heat conduction can be incorporated (Figures 4.23 and 4.24).

Three methods are basically used for this purpose: streamline curvature SLC methods (or SCM) (Novak 1967), matrix throughflow methods MTFM (Marsh 1968), and streamline throughflow methods (STFM) (Von Backström and Rows 1993). SCM allows simulating individual streamlines, along which flow properties are conserved, and this makes the methods relatively easy to implement and reasonably accurate. Conversely, a

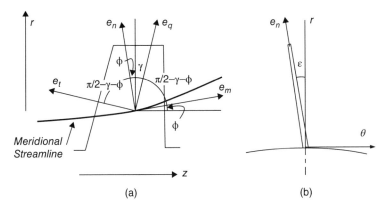

Figure 4.22 Coordinate system. (a) Unit vectors on meridional plane and (b) the angle of lean in a view along the axis. Source: Reprinted with permission from Korpela (2012) (figure A1 from original source), John Wiley & Sons © 2012.

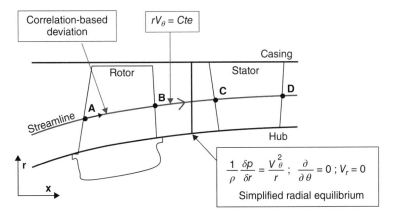

Figure 4.23 Domain sketch for throughflow calculations.

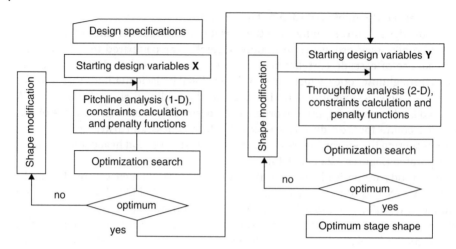

Figure 4.24 Optimization procedure proposed in Massardo et al. (1990). Source: Reproduced with permission from Massardo et al. (1990) (figure 1 from original source), Copyright © 1990 by American Society of Mechanical Engineers.

fixed geometrical grid is used in MTFM such that conservation properties along each streamline cannot hold. However, the adoption of a fixed calculation grid makes MTFM numerically more stable compared to SCM. Finally, STFM consists of a hybrid approach that conjugates benefits from SCM and MTFM.

4.2.2.4 Streamline Curvature Methods

The basic theory of SLC throughflow calculations has been described by many authors, particularly by Novak and Hearsey (1977). The basis of all throughflow methods is to obtain a solution for an axisymmetric flow and this may be pursued by either averaging all flow properties in the circumferential direction or by solving the axisymmetric flow equations on a mean blade-to-blade stream surface, the same set of equations being achieved in both ways, as reported by Denton (1978).

From the assumption of axial symmetry, it is possible to define a series of meridional stream surfaces as surfaces of revolution onto which the flow is assumed to lye all the way through the machine. The principle of the SLC method is to write the equations of motion along lines, known as quasi-orthogonals, or QOs (q), that are roughly perpendicular to these stream surfaces in terms of the curvature of the surfaces in the meridional plane.

What is required is an equation for pressure or any equivalent property gradients in the spanwise direction. This may be obtained from the equations of motion as described by Korpela (2012).

With reference to Figure 4.22, representing a mean stream surface on the meridional plane (a) and a view perpendicular to the rotational axis of the machine (b), the following expression for the steady-state flow acceleration can be derived:

$$\vec{a} = \hat{e}_m \, V_m \frac{\partial V_m}{\partial m} - \hat{e}_n \frac{V_m^2}{r_c} + \hat{e}_\theta \, V_m \frac{\partial V_\theta}{\partial m} - \hat{e}_r \frac{V_\theta^2}{r} \tag{4.55}$$

where $r_c = \partial m / \partial \phi$ is the radius of curvature of a generic streamline measured on the meridional surface (positive when, referred to the z-axis, the streamline is concave). The direction of the unit vector $\hat{e}_n$, which lies in the mean hub-casing stream surface inclined to the radial at an angle ε, known as angle of lean, is perpendicular to the direction of vector $\hat{e}_m$ on the meridional surface so that $(\hat{e}_n, \hat{e}_\theta, \hat{e}_m)$ form a right-handed triple. This is obtained by rotation by the angle ϕ about the axis of $\hat{e}_\theta$, hence the n direction accords with the radial coordinate and the m direction coincides with the axial direction when $\phi = 0$. It is worth noting that unit vector $\hat{e}_q$, lying on the meridional plane, represents the so-called quasi-orthogonal whose direction is chosen in advance roughly perpendicular to the streamlines and does not change during the calculation.

When rewritten for the triple $(\hat{e}_q, \hat{e}_t, \hat{e}_\theta)$, the preceding equation becomes the following, decomposed along the preceding unit vectors, which is very useful since the triple lies on the plane containing a blade with its lean angle:

$$a_q = \sin(\gamma + \phi) \, V_m \frac{\partial V_m}{\partial m} - \cos(\gamma + \phi) \frac{V_m^2}{r_c} - \frac{V_\theta^2}{r} \cos\gamma$$

$$a_t = \cos(\gamma + \phi) \, V_m \frac{\partial V_m}{\partial m} + \sin(\gamma + \phi) \frac{V_m^2}{r_c} - \frac{V_\theta^2}{r} \sin\gamma$$

$$a_\theta = V_m \frac{\partial V_\theta}{\partial m} \tag{4.56}$$

These components can be now inserted into the Euler equation for inviscid flows:

$$\vec{a} = -\frac{1}{\rho} \nabla p + \frac{\vec{F}}{\rho} \tag{4.57}$$

which together with the first law of thermodynamics and the definition of total enthalpy differentiated along the q direction:

$$T\frac{\partial s}{\partial q} = \frac{\partial h}{\partial q} - \frac{1}{\rho}\frac{\partial p}{\partial q}, \quad \frac{\partial h}{\partial q} = \frac{\partial h_0}{\partial q} - V_m \frac{\partial V_m}{\partial q} - V_\theta \frac{\partial V_\theta}{\partial q} \tag{4.58}$$

producing:

$$\frac{1}{2}\frac{\partial V_m^2}{\partial q} = \frac{\partial h_0}{\partial q} - T\frac{\partial s}{\partial q} + \sin(\gamma + \varphi) V_m \frac{\partial V_m}{\partial m} + \cos(\gamma + \varphi) \frac{V_m^2}{r_c} +$$

$$- \frac{1}{2r^2}\frac{\partial(r^2 V_\theta^2)}{\partial q} + \frac{V_m}{r}\frac{\partial(r V_\theta)}{\partial m} \tan\epsilon \tag{4.59}$$

This equation is the RE equation and represents the basis of all SLC calculation methods. This equation must be solved together with the continuity equation:

$$\int_{hub}^{casing} \rho V_m \cos(\gamma + \phi) w \, dq = \frac{\dot{m}}{N} \tag{4.60}$$

where $\dot{m}$ is the total mass flow rate, N the number of blades, and $w = 2\pi r B / N$ the portion of stream surface between adjacent blades, with B a measure of the blockage that would be equal to unity in an ideal flow. B is strongly affected by the boundary layer displacement on the annulus walls and blades, and is also reduced if the flow is not uniform in the circumferential direction, θ.

According to Denton (1978), in duct regions the distributions of total enthalpy, entropy, and angular momentum along the quasi-orthogonals is obtained from the conservation of these quantities along stream surfaces. Within blade rows, V_θ can be obtained from V_m, from the imposed flow directions and blade rotation, and from the blade geometry using correlations or blade-to-blade calculations. In stationary blade passages or outside blade rows the stagnation enthalpy h_0 is conserved along streamlines and can be calculated from the Euler equation; in moving blades it is the rothalpy $I = h + W^2/2 - U^2/2$ that is conserved. Entropy changes can be obtained from empirical loss correlations. The fluid density, needed when applying the continuity equation, may be obtained from the equation of state of the fluid once the enthalpy and entropy have been determined.

The method of solving the SLC equation is rather specialized and here below an example is illustrated, as reported by Cumpsty (1989):

1. choose the quasi-orthogonal positions;
2. guess the streamline shape in the meridional plane and evaluate streamline curvature and streamtube contraction at intersections with quasi-orthogonals;
3. guess meridional velocity V_m at each intersection of quasi-orthogonal and streamline and guess flow properties along the first quasi-orthogonal;
4. use blade-to-blade calculation or correlation with specified geometry and flow properties estimates to calculate flow outlet direction and loss, then calculate V and p_0 along the quasi-orthogonal (Figure 4.23);
5. evaluate terms on the right-hand side of SLC Eq. 2.3 starting from the first quasi-orthogonal, using current estimate for shape of meridional streamlines;
6. integrate $d(V_m^2)/dq$ along the quasi-orthogonal to get V_m with an arbitrary or guessed constant;
7. calculate overall mass flow rate from continuity Eq. 2.4 and adjust constant in predicted V_m distribution to get prescribed overall mass flow, then return to step 6 unless no adjustment needed, in which case go to 8;
8. integrate V_m to find new locations of meridional streamlines along the quasi-orthogonal for correct mass flow between them and store this information;
9. move to next quasi-orthogonal and repeat steps 4 to 8, then after last quasi-orthogonal go to 10;
10. allow intersection of streamlines with quasi-orthogonals to move toward the new position stored in step 8 but use relaxation factor to ensure stability, obtaining a new streamline shape and curvature;
11. go to step 5 unless movement required of streamlines is less than a convergence threshold, that is, a meridional solution is converged, in which case go to step 12;
12. obtain final results.

These methods have recently been made more realistic by taking account of end-wall effects and spanwise mixing by four aerodynamic mechanisms: turbulent diffusion, turbulent convection by secondary flows, spanwise migration of airfoil boundary layer fluid, and spanwise convection of fluid in blade wakes (Dunham 1997). Other remarkable results consist of incorporating a throughflow code into a Navier–Stokes solver for shortening the calculation phase (Sturmayr and Hirsch 1999).

A compressor map can be obtained showing very high accuracy using throughflow codes. Notable developments in the design methodologies have been obtained using such codes. Among others, Massardo et al. (1990) developed a method for the design optimization of an axial-flow compressor stage. The procedure allowed for optimization of the complete radial distribution of the geometry, being the objective function obtained using a throughflow calculation. Also, some examples were shown on the possibility to use the procedure outlined before both for redesign and the complete design of axial-flow compressor stages (Figure 4.24).

Howard and Gallimore (1993) described a viscous throughflow method for the design of an axial compressor including a meridional velocity defects in the end-wall region.

A remarkable application of a multi-objective evolutionary design optimization based on SLC throughflow code has been given by Oyama and Liou (2002a,b) for a four-stage compressor (Figure 4.25, left). In Figure 4.25 (right), Pareto optimal solutions are plotted revealing a substantial supremacy compared to the baseline compressor. Also, the complete span distribution of design variables was obtained, for example, the solidity of a blade that maximized one objective (see, for example, Figure 4.25).

4.2.2.5 Advanced Cascade Design Techniques (2D/Quasi-3D)

A great benefit in compressor design for maximum performance can be derived from advanced 2D/quasi-3D cascade aerodynamic design optimization using both direct

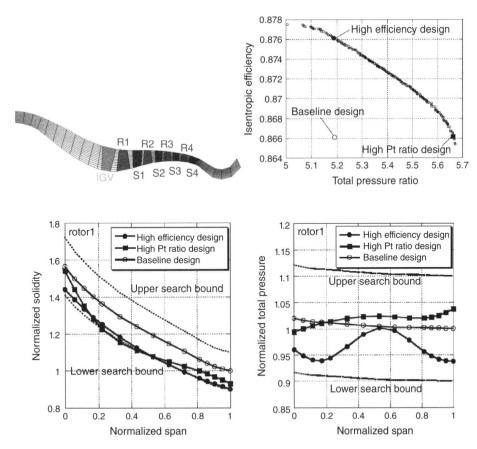

Figure 4.25 Throughflow optimization of a multistage compressor. Source: Reproduced with permission from Oyama and Liou (2002b) (figures 6, 7, 8, and 12 from original source), Copyright © 2002 by the American Institute of Aeronautics and Astronautics, Inc.

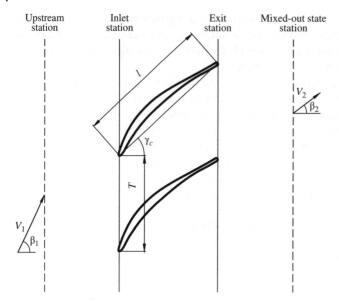

Figure 4.26 Two-dimensional cascade geometry for axial flow compressors. Source: Reproduced with permission from Benini and Toffolo (2002) (figure 1, figure 2, figure 3, figure 7 from original source). Copyright © 2004 by the American Institute of Aeronautics and Astronautics, Inc.

and indirect methods. In direct methods, a blade-to-blade geometry (Figure 4.26) is first established and subsequently analyzed using available flow solvers. Then, shape modifications take place and resulting geometries evaluated until an acceptable or even optimal configuration is found. Iterative methods perfectly suit this purpose, so that optimization loops are often used in the framework of this approach.

In a compressor cascade, the pressure rise is achieved by enthalpy exchange from the dynamic head (relative or absolute, according to the type of motion) and can be calculated using the following non-dimensional pressure ratio:

$$PR = \frac{\bar{p}_2}{p_1} \tag{4.61}$$

where the subscripts 1 and 2 refer to flow condition upstream and downstream of the cascade (Figure 4.26). On the other hand, the total pressure loss coefficient can be defined using the following equation:

$$\omega = \frac{p_{02}^{\,is} - \bar{p}_{02}}{p_{01} - p_1} \tag{4.62}$$

where the superscript *is* denotes the final isentropic state of the thermodynamic process and *o* refers to total quantities. In the preceding equations, the actual static pressure downstream of the cascade is calculated according to the mixed-out state hypothesis (Drela and Youngren 1995). According to this hypothesis, the implicit state of the flow in an arbitrary plane downstream of the cascade is obtained analytically from the known stagnation conditions, mass flow rate, and angular momentum:

$$\bar{\rho}\,\bar{u}Tbr = \int \rho ubrd\theta = \dot{m}$$

$$(\overline{\rho u}^2 + \bar{p})Tbr = \int (\rho u^2 + p)brd\theta = \int (u + p/(\rho u))d\dot{m} - \rho_e V_e u_e \Theta b + p_e \delta^* bV_e/u_e$$

$$\bar{\rho}\,\bar{u}\,\bar{v}Tbr = \int \rho uvbr\ d\theta = \int vd\dot{m} - \rho_e V_e v_e \Theta b \qquad (4.63)$$

where T is the angular pitch, u,v are the axial and tangential velocity components and $V = \sqrt{u^2 + v^2}$ is the flow speed. Θ and δ^* are the momentum thickness and the displacement thickness of the boundary layer at the cascade exit plane (denoted by the subscript e). Finally, b is the elementary streamtube height (see Figure 4.27). The mixed-out quantities in Eq. (4.3) can thus be used to calculate the Mach number and the total pressure at the station downstream:

$$\overline{M}_2^{\,2} = \frac{\bar{p}_2}{\gamma \bar{p}_2}\left(\bar{u}_2^{\,2} + \bar{v}_2^{\,2}\right)$$

$$\bar{p}_{02} = \bar{p}_2\left(1 + \frac{\gamma - 1}{2}\overline{M}_2^{\,2}\right)^{\frac{\gamma}{\gamma-1}} \qquad (4.64)$$

The determination of the maximum pressure rise that can be achieved in a compressor cascade is vital in the design of high-performance axial-flow compressors. As the pressure rise increases for a given inlet flow angle, the profile tends to become more cambered or has higher incidence angles, and thus a suction side more exposed to flow separation due to the adverse pressure gradients. In these circumstances, the total pressure losses generated inside the boundary layer, in the wake, and in separated flow regions become remarkable and they quickly convect in the mean flow. Moreover, the high-pressure rise associated with large turning at high inlet angles promotes turbulent separation so that the stall angle of attack moves progressively closer to the design value when section camber increases. Therefore, both pressure rise and total pressure loss are linked in such a way that the improvements obtained in one of the two significantly affect the other; in other words, a conflict appears between the two objectives of the design so that a trade-off between pressure rise and efficiency is unavoidable.

4.2.2.6 Geometry Definition and Parameterization

Using the Wu's S1 streamsurface concept, a generic compressor cascade geometry can be defined in a quasi-3D way using the well-known conformal mapping of coordinates, as seen in Section 4.1 (Figure 4.27):

$$m' = \int \frac{dm}{r} = \int \frac{\sqrt{dr^2 + dz^2}}{r} \qquad (4.65)$$

An option to handle such a problem is to optimize airfoil geometry by parameterizing its shape using Bézier parametric curves. A valuable choice is to use two Bézier curves (one for the pressure side and one for the suction side). $N + 1$ control points constituting the Bézier polygon define each curve. Therefore, the general expression for the Cartesian

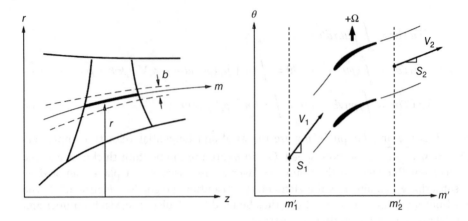

Figure 4.27 Stream surface definition. Source: From Drela and Youngren (2008).

coordinates of pressure side (abbreviated to *PS*) and suction side (abbreviated to *SS*) is:

$$
\begin{Bmatrix} x(t) \\ y_{PS}(t) \\ y_{ss}(t) \end{Bmatrix} = \sum_{i=0}^{n} C_{n,i} t^i (1-t)^{n=1} \begin{Bmatrix} x(i) \\ y_{PS}(i) \\ y_{SS}(i) \end{Bmatrix}
\tag{4.66}
$$

where $t \in [0, 1]$ is the non-dimensional parameter of each curve and $C_{n,i}$ are coefficients defined as:

$$
C_{n,t} = \frac{n!}{i!(n-i)!}
\tag{4.67}
$$

The control points' coordinates on airfoil contour are defined by the following equations:

$$
x(i) = \begin{cases} 0, & \text{if } i = 0; \\ \dfrac{i-1}{n-2}, & \text{if } 0 < i < n; \\ 1, & \text{if } i = n; \end{cases}
\tag{4.68}
$$

$$
(y_{PS}(i), y_{SS}(i)) = \begin{cases} 0, & \text{if } i = 0 \text{ or } i = n; \\ y_{CL}(i) \mp \delta(i), & \text{if } 1 < i < n-1; \\ \delta(i), & \text{if } i = 1 \text{ or } i = n-1; \end{cases}
\tag{4.69}
$$

where $y_{CL}(i)$ and $\delta(i)$ denote the i-th components of the "primary" control points; that is, the parameter set of the optimization for profile geometry (Figure 4.28). This method is inspired by the well-known practice of combining a camber line with a thickness distribution superimposed to the camber line. Such a method is effective in avoiding the creation of useless profiles that would come from the intersection of the pressure and the suction surface if these were defined separately. The leading edge and the trailing edge curvatures are described by control points lying on the $x = 0$ and $x = 1$ axes, respectively. Using this method, the direction of the camber line at the leading and the trailing edge

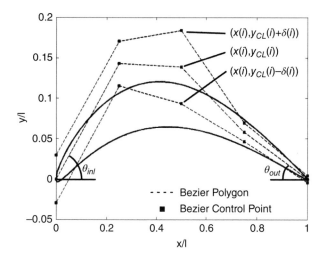

Figure 4.28 Geometry parameterization of an airfoil using Bézier curves; squares represent control points of Bézier curves. Source: Reproduced with permission from Benini and Toffolo (2002) (figure 2 from original source) Copyright © 2004 by the American Institute of Aeronautics and Astronautics, Inc.

can be well approximated by:

$$\theta_{inl} = \frac{y_{CL}(1)}{x(2)} \quad \theta_{out} = \frac{y_{CL}(n-1)}{x(n) - x(n-1)} \tag{4.70}$$

so the incidence angle $\alpha_1 - \beta_1 - \theta_{inl}$ can be easily specified.

In order to maximize pressure ratio and minimize total pressure losses across the cascade for a given inlet Mach number, inlet, and outlet flow inclinations must be established first. Next, a constraint on maximum allowable total pressure losses over a pre-defined operating range of the cascade can be taken into account leading to a multi-point optimization problem, for example, $\omega_i/\omega^* \leq 2 \ \forall i = 1,\ldots,m$, with ω_i being the total pressure loss coefficient of the cascade at point i and ω^* the reference total pressure loss coefficient (Figure 4.29).

A constraint on the operating range has an incisive impact both on fluid dynamic and structural aspects. It is primarily useful for determining an acceptable stall/surge margin of the compressor. Moreover, it controls indirectly the shape of the leading edge by making it sufficiently rounded and regular, and addresses the geometry search toward profiles having adequate structural strength. Such condition is definitely useful in governing the maximum thickness of the profile without explicit constraints that would limit the exploration of non-conventional solutions within the design space.

When an accurate definition of the baseline airfoil geometry is available, that is, there is no need to reconstruct it via a parametric representation, a valuable alternative to simple Bézier curves is to use a superimposition of a displacement field to the baseline geometry:

$$\mathbf{P}_{mod}(u) = \mathbf{P}_{base}(u) + \mathbf{P}_{displ}(u). \tag{4.71}$$

where u is a parameter.

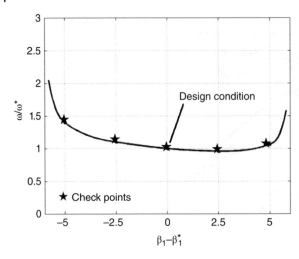

Figure 4.29 Total pressure loss as function of inlet flow angle at the design point (dots represent the sampled points considered in constraint evaluation). Source: Reproduced with permission from Benini and Toffolo (2002) (figure 3 from original source). Copyright © 2004 by the American Institute of Aeronautics and Astronautics, Inc.

Such a displacement field can be properly defined by using a generalized form of the Bézier curves, for example, Basis Splines, otherwise referred to as B-spline curves. The reader is referred to the work of Mortenson (1997) for a detailed explanation of B-spline curves and their numeric implementation.

B-splines do have several advantages over simple Bézier curves due to the fact that the position of the control points does not influence the curve globally and that the curve degree does not depend on the number of control points. Furthermore, enforcing C2 continuity (i.e., identical tangents and identical curvature at the curve knots or breakpoints), a property that is usually appreciated in blade design, in simple Bézier curves is only possible at the price of having the control points dependent on each other. On the other hand, B-Splines do have C2 continuity by definition if properly defined, as shown later on.

$$\mathbf{P}_{displ}(u) = \sum_{i=0}^{n} N_{i,p}(u)\mathbf{P}_i, \tag{4.72}$$

The displacement field is defined through a B-spline as follows:
where

- $\mathbf{P}_i$ are the $n+1$ control points.
- $N_{i,p}(u)$'s are B-spline basis functions polynomials of degree p defined on a knot vector $\mathbf{U} = (u_0, \ldots, u_m)$ having $m+1$ knots in ascending order. Note that u can be defined either over the interval $[0,1]$ or rescaled as preferred.
- n, m and p must be such that $m = n + p + 1$. In particular, a B-spline curve of degree p with $n+1$ control points is defined by $n+p+2$ knots, that is, $u_0, u_1, \ldots, u_{n+p+1}$. On the other hand, if a knot vector of $m+1$ knots and $n+1$ control points are given, the degree of the B-spline curve is $p = m - n - 1$. The curve point corresponding to a knot u_i, $C(u_i)$, is called a knot point. Hence, B-spline curve is divided into curve segments, each of which is defined on a knot span.

$$N_{i,1} \begin{bmatrix} = 1 & \text{if } u_i \leq u \leq u_{i+1} \\ = 0 & \text{otherwise} \end{bmatrix}. \tag{4.73}$$

$$N_{i,p}(u) = \frac{(u - u_i)N_{i,p-1}(u)}{u_{i+p-1} - u_i} + \frac{(u_{i+p} - u)N_{i+1,p-1}(u)}{u_{i+p} - u_{i+1}} \qquad (4.74)$$

- The basis functions are defined in the following way:
- u_i are the knot values that relate the parametric variable u to the control points $\mathbf{P}_i$. If locking (clamping) the curve is desired so that it is tangent to the first and the last segments at the first and last control points, respectively, as a Bézier curve, the first knot and the last knot must be of multiplicity $p + 1$.

The degree of a B-spline basis function is an input, while the degree of a Bézier basis function depends on the number of control points. If a change in the shape of a B-spline curve is desired, one can modify one or more of these control parameters: the positions of control points, the positions of knots, and the degree of the curve. It is relatively easy to demonstrate that if each basis function is of k-th order then C2 continuity is assured when $p \geq 3$.

4.2.2.7 Flow Solvers

4.2.2.7.1 2D, Quasi-3D Inviscid-Viscous Codes

A valuable option to perform rapid 2D and quasi-3D blade-to-blade computations in compressor cascades is to use inviscid-viscous solvers, an example of which has been implemented by Giles and Drela (1987). These codes have proven to give satisfactory results for compressible flows up to highly subsonic incoming conditions ($M < 0.85$), as well as for attached or moderately separated cascade flows. However, when incoming supersonic flows are dealt with, some issues may occur due to code instabilities although shock-capturing schemes are often used and successfully implemented in such formulations.

In such codes, a two-zone approach is used: the viscous flow is solved indirectly based an equivalent inviscid flow, which is guessed outside a displacement streamline that includes the boundary layer. The equivalent inviscid flow is defined as locally irrotational and containing all the mass flow; the viscous effects are modeled using the mass flow defect. The inner boundary of the equivalent inviscid flow is displaced outward from the wall by the displacement thickness of the boundary layer.

The outer inviscid compressible flow is solved using the steady-state integral form of Euler equations applied to a closed control volume C where ds is the infinitesimal segment on the C boundary:

$$\oint \rho \mathbf{c} \cdot \mathbf{n} \; ds = 0 \quad (continuity) \qquad \oint [\rho(\mathbf{c} \cdot \mathbf{n}) \; \mathbf{c} + p\mathbf{n}] \; ds = 0 \quad (momentum)$$

$$\oint \rho \mathbf{c} \cdot \mathbf{n} h_0 ds = 0 \quad (energy) \qquad (4.75)$$

A space discretization is applied to these equations, which are solved on a grid. This grid is usually organized such that its cells feature at least two opposite faces lie on streamlines, in order to have zero mass flux across each of them. In such a way, the grid itself is part of the solution and changes dynamically from an initial trial to a final converged one (Giles and Drela 1987; Drela and Youngren 1995). A modified linearized Newton iteration including a Gaussian elimination block is used to obtain an approximation of the flow solution at each iteration. At each code iteration, both the flow field variable and the streamline position are updated using relaxation techniques.

The boundary layer flow is solved using an integral boundary layer method based on the Prandtl differential equations for adiabatic, laminar/turbulent flows, expressed in local streamwise and normal wall coordinates:

$$\frac{\partial(\rho u)}{\partial \xi} + \frac{\partial(\rho v)}{\partial \eta} = 0 \quad \text{(continuity)}$$

$$\rho u \frac{\partial u}{\partial \xi} + \rho v \frac{\partial v}{\partial \eta} = \rho_e u_e \frac{du_e}{d\xi} + \frac{\partial \tau}{\partial \eta} \text{(momentum)}$$

(4.76)

where τ is the local shear stress (including Reynolds stresses):

$$\tau = \mu \frac{\partial u}{\partial \eta} - \overline{\rho u' v'}$$

(4.77)

And "e" denotes the edge quantities at the boundary layer thickness.

A well-known integral solution to the momentum equation is due to von Kármán:

$$\frac{\xi}{\theta} \frac{d\theta}{d\xi} = \frac{\xi}{2} \frac{C_f}{2} - \left(\frac{\delta^*}{\theta} + 2 - M_e^2\right) \frac{\xi}{u_e} \frac{du_e}{d\xi}$$

(4.78)

where δ^*, θ, and C_f are the displacement thickness, momentum thickness, and the skin friction coefficient, respectively:

$$\theta = \int_o^\infty \left(1 - \frac{u}{u_e}\right) \frac{\rho u}{\rho u_e} d\eta \quad \delta^* = \int_o^\infty \left(1 - \frac{\rho u}{\rho u_e}\right) d\eta \quad C_f = \frac{2}{\rho_e u_e^2} \tau_w$$

(4.79)

In inviscid-viscous codes such as the one proposed by Drela and Youngren (1995), an additional equation is used that improves accuracy over the simpler Thwaites' solution. Such a solution is derived directly from the momentum equation, pre-multiplied by the u-velocity component and then integrated to obtain a kinetic energy integral equation:

$$\frac{\xi}{\theta^*} \frac{d\theta^*}{d\xi} = \frac{\xi}{\theta^*} 2C_D - \left(\frac{2\delta^{**}}{\theta^*} + 3 - M_e^2\right) \frac{\xi}{u_e} \frac{du_e}{d\xi}$$

(4.80)

where:

$$\theta^* = \int_o^\infty \left(1 - \left(\frac{u}{u_e}\right)^2\right) \frac{\rho u}{\rho u_e} d\eta \quad \delta^{**} = \int_o^\infty \left(1 - \frac{\rho}{\rho_e}\right) \frac{u}{u_e} d\eta$$

$$C_D = \frac{1}{\rho_e u_e^3} \int_o^\infty \tau \frac{\partial u}{\partial \eta} d\eta$$

(4.81)

Edge quantities are initially known from the Euler solution outside the boundary layer. Laminar closure allows the solution of the laminar zone of the boundary layer. Wall closure is provided using empirical Falkner–Skan family of velocity profiles (Giles and Drela 1987). Wall turbulent closure is provided using a two-parameter turbulent mean velocity profile. To this purpose, many options are available today starting from profiles proposed by Swafford and by Giles and Drela (1987). In free wakes, the integral equations remain valid as long as wall skin coefficient is discarded. Since it is almost impossible for laminar free wakes to take place downstream of an airfoil (inflection in the velocity profiles promote turbulent dissipation), only turbulent free wake closure relations need to be used (therefore, $C_f = 0$ in turbulent integral boundary layer equations).

Most of these codes use a transition criterion (Drela and Youngren 1995) to predict the transition region of the boundary layer from laminar to turbulent. Usually, some variants of the well-known e^n methods are used for this purpose. Such methods provide equations for the prediction of growth and decay of infinitesimal wave-like disturbances that anticipate free transition. Next, the transition is assumed to occur when the cumulative amplitude of disturbances reaches a certain value, which is often chosen as a natural power of the exponential function (e.g., $e^{10} \approx 22\,000$).

The outer flow is coupled with the boundary layer one through the edge velocity and density.

The coupling usually occurs as follows:

- First, the inviscid surface streamline is displaced from the wall by a distance Δn equal to a postulated boundary layer displacement thickness δ^* (Figure 4.30). In wakes, the pertinent thickness Δn is imposed to be the sum of δ^* values at profile TE on both the suction and pressure side and propagated downstream according to the momentum equation.
- Next, the local inviscid velocity (q in Figure 4.30) components must be constrained to the edge velocity, for example using $u_{e2} = 0.5(u_i + u_{i-1})$ or slightly different formulations useful in the case of mild separated flows.
- Boundary layer momentum and kinetic energy equations are linearized and coupled with the solid wall boundary conditions derived from the Euler equations. The coupled system of equations is solved using a Newton's scheme.
- The Newton's iteration provides updated values of all the flow quantities and integral properties of the boundary layer.
- The process is repeated until convergence is obtained. To facilitate convergence, under-relaxation factors can be adopted during quantity updating.

Because the inviscid and viscous solutions are fully coupled and solved simultaneously, flows involving separation can be calculated. However, it can handle only moderate stall phenomena since, for massive separations, the iteration procedure does not converge. The influence of the stream tube height is considered and for local supersonic regions the "artificial viscosity" formulation is implemented. Flow quantities of interest, such as exit flow angle, pressure rise, and total pressure loss, are

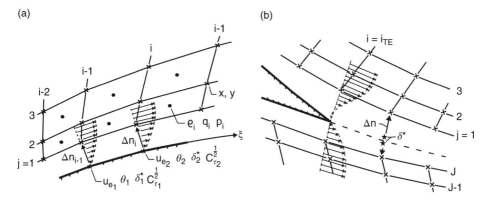

Figure 4.30 Example of coupling inviscid and viscous solutions in wall boundary layers (a) and wakes (b). Source: Reprinted with permission from Drela (1986). © 1986.

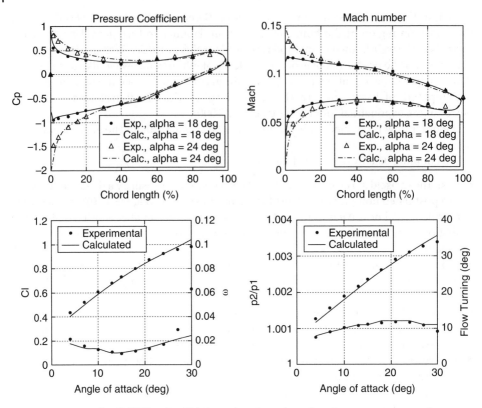

Figure 4.31 Example of MISES code validation subsonic computations in compressor cascades. Source: Reproduced with permission from Benini and Toffolo (2002) (figure 7 from original source). Copyright © 2004 by the American Institute of Aeronautics and Astronautics, Inc.

calculated using the equations of mass, momentum, and energy conservation. Structured grids (I-type and H-type) with elliptic smoothing are usually employed for space discretization.

As an example of validation, Figure 4.31 shows pressure and Mach number distribution on pressure and suction side of a NACA 65 cascade of airfoils at low speed with $\beta_1 = 30°$, $\sigma = 1$, and $AVDR = 1.0$, for 18 and 24° angle of attack obtained using the MISES code (Benini and Toffolo 2002). The computed values exhibit good agreement with experimental data published by Herrig et al. (1957). The values of pressure ratio, total pressure losses, and flow turning were compared with experimental ones as well for a large number of incidence angles. The results obtained are regarded as satisfactory in all the conditions explored during the validation.

4.2.2.8 3D Methods

4.2.2.8.1 Direct Methods
Advanced optimization techniques can be of great help in the design of 3D compressor blades when direct methods are used. These are usually very expensive procedures in terms of computational cost such that they can be profitably used in the final stages of the design when a good starting solution, obtained using a combination of 1D and/or 2D

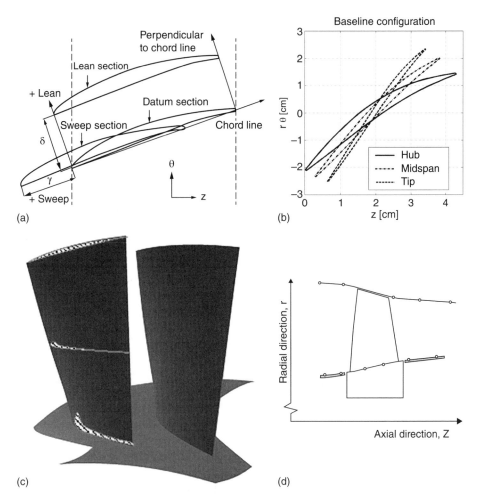

Figure 4.32 Definition of geometric sweep and lean in a transonic compressor rotor blade.

methods, is already available. Moreover, large computational resources are necessary to obtain results within reasonable industrial times.

3D optimization methods are particularly useful in defining the 3D stacking line of both rotor and stator blades in order to obtain maximum compressor performance. Typical objectives include maximum efficiency, maximum pressure ratio per single stage, and maximum operating range between surge and stall. It is well-known that 3D flows occur in a compressor blade row, which includes both a viscous and non-viscous effect that ultimately leads to 3D losses and entropy generation that negatively influence the overall efficiency. Actually, the application of aerodynamic sweep and lean on rotor and stator blades is one of the most significant technological evolutions to improve the aerodynamic behavior of compressor stages. Aerodynamic sweep and lean are represented in Figure 4.32: sweep is defined as the movement of blade sections along the local chord line while lean is the change in the perpendicular direction (Figures 4.33 and 4.34).

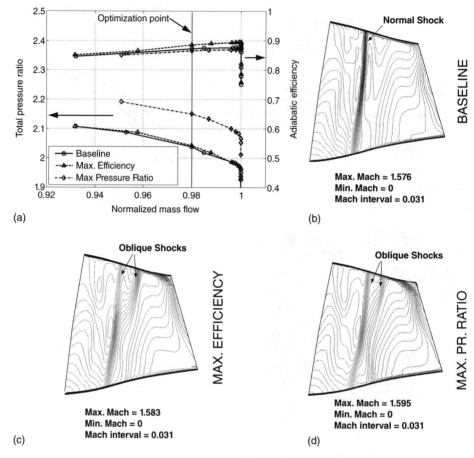

Figure 4.33 Performance map and Mach number contours of baseline and optimized compressor configurations. Source: Reproduced with permission from Benini (2004) (figure 10 from original source). Copyright © 2004 by the American Institute of Aeronautics and Astronautics, Inc.

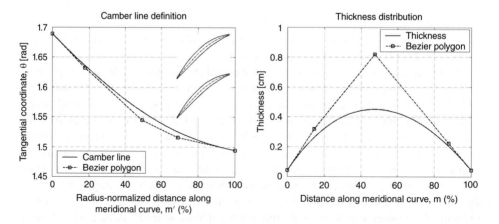

Figure 4.34 Definition of camber line and thickness distribution in a compressor airfoil. Source: Reproduced with permission from Benini (2004) (figures 3 and 7 from original source). Copyright © 2004 by the American Institute of Aeronautics and Astronautics, Inc.

Sweep and lean do have effects on the blade circulation and losses. Blade sweep influences the loading on the blade near the walls. In particular, blade loading in the front area of the tip region can be reduced in subsonic compressor blades by means of forward sweep. This is helpful in reducing the sensitivity to modifications in incidence and the intensity of the tip leakage flows. In transonic compressor rotors, forward sweep moves the shock downstream near the casing, and this helps to delay the vortex breakdown leading to better compressor stability. On the other hand, some research has shown appreciable improvements in the overall rotor performance when blades are properly curved in the tangential direction using lean. In particular, skewing the blade toward the direction of rotation has demonstrated the possibility to increase overall rotor efficiency. The starting point consists of finding a suitable parameterization technique for geometry representation and manipulation.

4.2.2.8.2 *Optimization Involving Direct Methods*
An optimization technique combined with a 3D Navier–Stokes solver to find the optimum shape of a stator blade in an axial compressor was described by Lee and Kim (2000). For numerical optimization, searching direction was found by the steepest decent and conjugate direction methods, and the golden section method was used to determine optimum moving distance along the searching direction. The objective of optimization was to maximize efficiency. An optimum stacking line was also found to design a custom-tailored 3D blade for maximum efficiency with the other parameters fixed.

An example of advanced 3D design of industrial compressors blades was given by Sieverding et al. (2004), in which a parametric geometry method, a blade-to-blade solver, and an optimization technique (i.e., a GA) were linked together. Optimized compressor profiles showed an enlarged operating range compared to the baseline design featuring NACA65 profiles.

A multi-objective design optimization method for 3D compressor rotor blades was developed by Benini (2004) in which both isentropic efficiency and pressure ratio at the design point were maximized (Figure 4.33). 3D RANS equations and a multi-objective evolutionary algorithm were used for handling the optimization problem. Compressor blade shape was parameterized by means of three profiles along the span (hub, midspan, and tip), each of which was described by camber and thickness distributions, both defined using Bézier polynomials (Figure 4.34).

4.2.3 Radial Compressor Optimization

Centrifugal compressors are currently being designed to achieve high stage pressure ratios with relatively low mass flow rates. Advanced new designs are documented that show superior performance compared to solutions adopted in the past in terms of both pressure ratio and efficiency (Japikse 2000). Among these, compressors with the highest pressure ratios feature the lowest efficiency and, vice versa, configurations with the highest efficiency do have inevitably the lowest pressure ratios. In many cases the designer might be interested in knowing the best compromise between efficiency and pressure ratio that he/she may be able to achieve: the quality of a new aerodynamic design is thus measured with respect to more than one objective, that is, such a design problem is inherently of multi-objective nature. In this context, the use of evolutionary algorithms can be of considerable help since they adopt a population of candidate solutions to the optimization problem, and therefore are suitable to capture the entire set of trade-offs among multiple design objectives.

Applications of advanced design methods and optimal design techniques regarding centrifugal compressor components are well documented in the open literature. Al-Zubaidy (1990) coupled a quasi-3D flow analysis program with a deterministic optimization algorithm to ensure an acceptable diffusion along the flow path of a compressor impeller. Zangeneh et al. (1999) applied a 3D inverse design method with the aim of suppressing secondary flows and indirectly enhance compressor stability range. Yiu and Zangeneh (1998) developed a "hybrid" 3D inverse design method coupled with a direct approach using CFD to minimize the loss of impeller blades. More recently, Cosentino et al. (2001) used a GA and an artificial neural network to improve the efficiency of a centrifugal compressor impeller. Bonaiuti and Pediroda (2001) applied a GA to maximize the efficiency of a centrifugal compressor impeller and analyzed the influence of each of the most common design parameters on the objective of the optimization. Bonaiuti et al. (2002) developed an optimization technique for maximizing transonic compressor impeller efficiency using the design-of-experiment technique. Benini and Tourlidakis (2001) showed how a multi-objective evolutionary technique can be used to optimize a vaned diffuser of a centrifugal compressor. Zangeneh et al. (2002) applied a 3D inverse design method to improve the design of the vane geometry of a centrifugal compressor vaned diffuser.

In the following, 3D methods are described that are able to assist the designer of centrifugal compressor impellers in the accomplishment of multiple objectives with the presence of constraints.

4.2.3.1 3D Models

The role of 3D flows and viscous effects in centrifugal compressor blades is fundamental and must be taken properly into account in the design optimization to obtain maximum performance. To this purpose, CFD can provide very useful and detailed knowledge about the 3D flow field inside the vane passages. However, employing CFD to improve compressor performance is sometimes hard because it is difficult to know how to effectively extrapolate information from numerical results in order to modify the blade geometry toward the optimum solution. In the following, a fully 3D direct method is described that is able to assist the designer of centrifugal compressor impellers in the accomplishment of the required objectives with the presence of constraints.

Prior to a proper formulation of an optimization problem, a parametric shape model of both impeller and diffuser must be constructed as follows. In modern centrifugal compressor aerodynamic design, the number of geometrical parameters to be considered for optimization is usually very high and this makes the decision-making problem very hard to solve. The design is still characterized by overly long and repetitive iterations, even for an experienced designer with solid basis in physics and with the very best CFD technology available. In particular, the influence of all the design parameters on the overall compressor performance is almost impossible to be predicted in advance; furthermore, understanding how the parameters are linked together to determine the overall design value is a very difficult and multi-dimensional problem that requires advanced mathematical models. On the other hand, when more than one objective is to be considered in the design, problem complexity rises tremendously.

The 3D geometry of radial impellers is usually defined following the effective method introduced by Casey (1983). Such a method makes use of Bernstein–Bézier polynomial

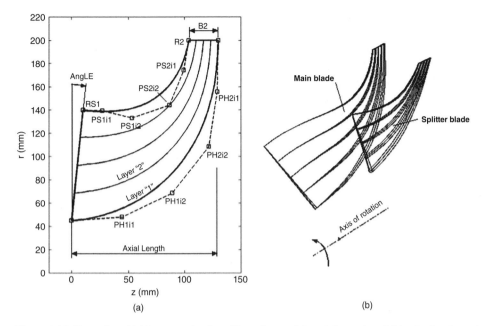

Figure 4.35 Examples of (a) Parameterization of impeller meridional channel, and (b) pair of main and splitter blades obtained using the parametric code.

patches of prescribed order to describe the geometrical shape of both meridional and blade surface channels.

In the meridional channel, fifth-order Bézier curves usually provide very accurate representations of the throughflow path geometry in the r-z plane of the impeller: usually, two limit curves are parameterized, that is, one for the hub layer and one for the shroud layer (Figure 4.35), each of which has the following expression:

$$\left\{ \begin{array}{c} r \\ z \end{array} \right\}_m = \sum_{k=0}^{5} B^5{}_k(t) \cdot P_k = \sum_{k=0}^{5} \binom{5}{k} t^k (1-t)^{5-k} \left\{ \begin{array}{c} r(k) \\ z(k) \end{array} \right\} \tag{4.82}$$

A total of five layers are typically used to complete the impeller parameterization in the meridional plane. Hub and shroud curves are given the name of layer "1" and layer "5," respectively. Intermediate layers 2–4 are defined in the meridional channel as equally spaced span fraction layers to allow detailed definition of 3D blade shape. The number of parameters is often adequate to explore a wide range of geometrical configurations and, at the same time, is suitable to reproduce meridional channels from data that is more readily available to the designer, such as radii of hub and shroud contours and axial and radial lengths of impeller sections.

In the blade surface channel, both Bézier and spline curves can be employed to describe basic blade definition parameters on a layer (i.e., on a trace of a blade-to-blade surface) as a function of meridional coordinates m and m'. This has the advantage of making the surfaces of compressor impeller easy to manipulate parametrically and gives continuous derivatives up to very high orders.

Impeller blade profiles are often parameterized using the sequence of operations as follows:

(i) *Profile camber line* is defined using fifth-order Bézier curves laying in the conformal plane m'-θ corresponding to each layer, so that a total of five camber lines are needed per single blade. Each curve identifies the distribution of profile camber line angle θ from leading edge to trailing edge (Figure 4.36a).

(ii) Blade thickness distribution is specified using fifth-order Bézier curves for each camber line as a function of the conformal meridional distance m' (Figure 4.36b) and superimposed to blade camber line to obtain actual profile coordinates. The leading edge has the shape of a circular arc or an ellipse, while the trailing edge shape is cut off. In most design cases, however, a constant thickness distribution with a circular leading-edge blade per layer and a linear spanwise thickness distribution from hub to shroud are adequate for both aerodynamic and structural purposes.

(iii) Splitter *blades*, when present, are blades positioned between main blades for additional flow control and for reducing main blade loading. In the procedure described here, splitter blades are "dependent" on the main blade for their angular, meridional, and thickness definitions. Following this, the splitter circumferential position is specified using an offset angular coordinate expressed as a fraction of angular pitch between main blades (from 0 to 1). Splitter camber line is simply derived from that of the main blade once a value of m'_{spl}/m', that is, a starting point in the meridional plane is specified. Splitter thickness distribution is obtained from that of the main blade; that is, points with the same radius on the two blade camber lines have the same thickness.

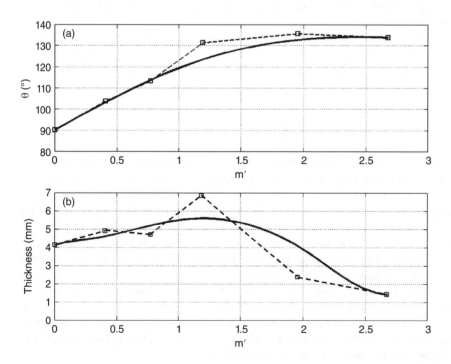

Figure 4.36 Examples of parameterization of blade camber line (a) and thickness distribution (b) using Bézier polynomials.

The procedure outlined here is able to generate pairs of very good blade shapes using 14–16 parameters. Figure 4.35b shows an example of a pair of blades (main and splitter) obtained using this method. The use of parametric Bézier curves makes the definition of blade shapes simple and straightforward, and makes it possible to obtain very regular and smooth surfaces that can be examined using aerodynamic and structural codes.

4.2.3.2 CFD Analysis

General agreement exists on the role that CFD will play in the development of effective design optimization techniques (Japikse 2000). A strong contribution come from the use of commercial codes, being so widespread both in academic and industrial organizations, provided that they are previously and rigorously validated.

A paper by Tsuei et al. (1999), reports the results of validation for two well-known commercial codes (Pushbutton CFD,[1] derived from Dawes' code BTOB3D, and NUMECA's FINE/Turbo[2]). The relevant outcome of this study is that most of the important effects in turbomachinery may be captured using a coarse grid of only 30 000 nodes. As a result, the use of relatively simple 3D CFD models for computing the flow in compressor blades is often pursued in optimization problems. CFD codes are usually validated against experimental results, for example, from the well-known Eckardt compressor impeller "0" (Eckardt 1975, 1987, 1979; Eckardt and Trültzsch 1977).

A single-block grid of $49 \times 30 \times 21 = 30\,870$ nodes can be employed for agile optimization runs (Figure 4.37c, d). A standard k-ε turbulence model can be used along with standard wall functions (Reynolds number $Re_W = 7.0 \times 10^6$ at shroud trailing edge and $Re_W = 1.0 \times 10^6$ at hub trailing edge). The parameter $y+$ is kept between 50 and 100 along the blade and end-wall regions. If the impeller is modeled without tip clearance some overprediction of efficiency can be expected. The CFD results are in good agreement with measured data for all the conditions explored. At the design rotational speed, computed results regarding total pressure ratio agree with measured data from −3% (choke) to +1% (surge). At 16 000 rpm, where more evident discrepancies are registered, computational accuracy is always in the range of −4% (choke) to +0.5 (surge), however, it is within the uncertainty level of experimental data. Regarding rotor efficiency, the code behaved quite well in all the conditions investigated, the maximum error being in the order of 0.5%. It is worth noting that the code captured the true efficiency pattern in the operating range. Validation tests were repeated using a grid of $100 \times 50 \times 20 = 100\,000$ nodes, with no noticeable accuracy improvement. Furthermore, computations were also performed using a k-ε turbulence model, with no practical benefit, and an increased computational time of 1.5 times. This demonstrates that, for design purposes, a commercial code with relatively coarse grid and no exotic numerical features is effective to support an optimization process.

Validation results using unstructured tetrahedral grid of 20 043 nodes are reported in Figure 4.37a and b. The CFD results are in good agreement with measured data for all the conditions explored. Results of the validation process, although being referred to a particular geometry, suggest that the performance of a compressor impeller can be captured with a relatively simple CFD model and a reasonably coarse grid.

---

1 Pushbutton CFD is a trademark of Concepts ETI, Inc.
2 FINE/Turbo is a trademark of NUMECA International, s.a.

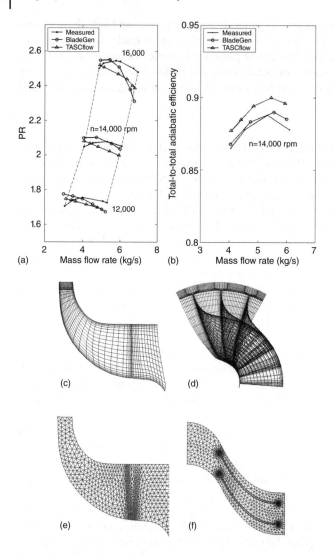

Figure 4.37 Experimental results versus computed predictions of pressure ratio (a) and isentropic efficiency (at nominal speed), (b) for Eckardt impeller "0." Structured grid (c–d). Unstructured tetrahedral grid (e–f).

4.2.3.3 Multi-Objective Optimization Problem and Results

The method described in the previous sections was used to obtain the Pareto Optimal Set (POS) that dominates the performance of the Eckardt impeller with respect to both efficiency and pressure ratio. Even though the Eckardt impeller may not be representative of some of the more advanced designs in use today, it provides a good basis of comparison and a solid base to interpret the results of the optimization. The Eckardt impeller is now called the "reference" impeller or the "original" design. A list of the decision variables and their respective range of variation is reported in Table 4.1. During the optimization process, the following boundary conditions were considered. At impeller inlet, the total pressure and total temperature were applied ($p_{01} = 101\ 325$ Pa,

Table 4.1 Design constraints on decision variables. Abbreviations refer to Figure 4.32.

Meridional section parameters				Blade-to-blade parameters			
Parameter name	Minimum	Maximum	No. of bits	Parameter name	Minimum	Maximum	No. of bits
RH1 (mm)	45	45	7	NumBlades	15	25	7
RS1 (mm)	140	140	7	BetaS1 (degree)	58	65	7
Angle (degree)	−5	15	7	Beta2 (degree)	−3	40	7
R2 (mm)	200	200	7	ThetaH2 (degree)	43.838	43.838	7
B2 (mm)	20	30	7	ThetaS2 (degree)	43.226	43.226	7
Axial length (mm)	112.7	150	7	Lean (degree)	−30	30	7
PH1i1_z (mm)	44.611	47.27	7	mH1 (%)	14.7	15.1	7
PH1i1_r (mm)	45.87	47.88	7		13	14.2	7
PH1i2_z (mm)	86.3	102.16	7	mH2 (%)	73	73.7	7
PH1i2_r (mm)	55.25	85.9	7		73.8	74.3	7
PH2i2_z (mm)	102.36	142.7	7	ThetaH1i2_m (degree)	28.7	31.1	7
PH2i2_r (mm)	98.13	106.81	7	ThetaH1i2_h (degree)	19.394	29.515	7
PH2i1_z (mm)	112.07	148.31	7	ThetaH2i2_m (degree)	44.1	54.2	7
PH2i1_r (mm)	152.7	163	7	ThetaH2i2_h (degree)	16.675	43.788	7
PS1i1_z (mm)	23.17	23.17	7	ThetaS1i2_m (degree)	30	30.2	7
PS1i1_r (mm)	140	140.04	7	ThetaS1i2_h (degree)	17.06	34.171	7
PS1i2_z (mm)	42.9	50.17	7	ThetaS2i2 m (degree)	46	51.6	7
PS1i2_r (mm)	142.8	145.15	7	ThetaS2i2_h (degree)	25.53	40.711	7
PS2i2_z (mm)	71.2	71.62	7	ThicknessH (mm)	3.5	3.5	7
PS2i2_r (mm)	142.4	152.3	7	ThicknessS (mm)	1.5	1.5	7
PS2i1_z (mm)	79.13	117.33	7	PitchFract	0.5	0.5	7
PS2i1 r (mm)	123	178.5	7	Lecutoff (%)	0	0	7

$T_{01} = 288.1$ K) and the flow was supposed to be swirl-free. Since the design operation of the original impeller is far from the choke condition, the mass flow rate was assigned at impeller exit ($\dot{m} = 5.31$ kg s^{-1}). The rotational speed was fixed at 14 000 rpm. In order to guarantee an acceptable operating range, a constraint on the actual throat area was applied and handled with a simple penalty function. In this optimization, the splitter blades were not considered.

In the optimization process, 10 individuals were used for a total of 120 generations. The optimization run resulted in a set of impellers with better overall performance at the design point. Figure 4.38 shows the final Pareto front obtained. In this figure, the performance of Eckardt impeller, registered numerically, is reported as well for reference. The right-hand-side of the Pareto front contains individuals with the highest pressure ratios; the left-hand side contains those with the highest efficiencies. It is worth noting that, for almost the same pressure ratio there is a solution belonging to the POS that has higher efficiency compared to the reference impeller. This impeller has been given the name of impeller "B"; the impeller with the highest efficiency has instead been given the name of

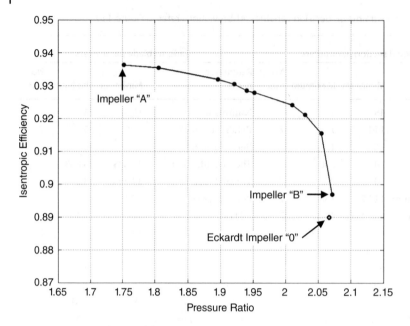

Figure 4.38 Pareto front (solid line) obtained after the optimization and original impeller design.

impeller "A" (these names are purely conventional and must not be confused with the impellers "A" and "B," which were tested by Eckardt). The meridional channel and the blade camber line distribution of these impellers are reported in Figure 4.39 parts a and b, respectively. In Figure 4.40, the solid models of the three impellers are shown.

The meridional geometry of impeller A is very similar to that of the reference impeller. However, due to the trailing edge backsweep, the pressure ratio is much lower. The leading edge is of forward-swept shape and the profile stacking is no longer purely radial: the blades are apparently leaned both at the leading and the trailing edges in the direction of rotation (+11° and +17° with respect to tangential direction, respectively). Blade incidence angles are not significantly changed after the optimization but the camber line curvature is appreciably decreased. The number of blades reduces to 18. Such modifications determine a significant change in the blade wet surface and loading.

Impeller "B" develops a compression ratio similar to that of the reference impeller (in fact, it has radially ending blades) but exhibits a higher efficiency. In spite of the increased number of blades (24 Vs. 20), this impeller takes advantage of the profile lean at both the leading and trailing edges (+10° and +13° with respect to tangential direction, respectively). The impeller has substantial forward-swept leading edges and a different meridional channel curvature. The curvature of blade profiles is mainly concentrated toward the trailing edge.

4.2.4 Turbines

Gas and steam expanders are broadly divided according to mean flow path configuration, that is, comprise both axial-flow and radial flow (both inflow and outflow) paths.

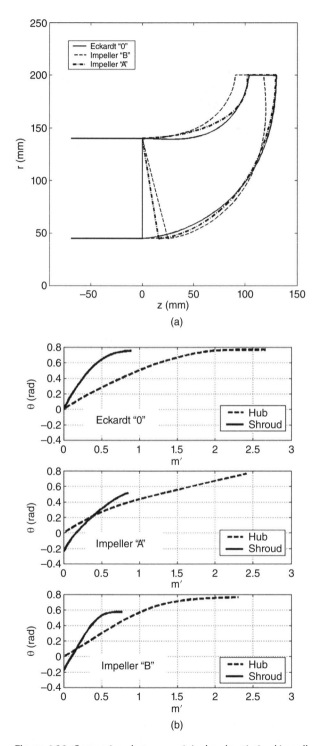

Figure 4.39 Comparison between original and optimized impellers. Meridional view (a) and camber line distribution (b).

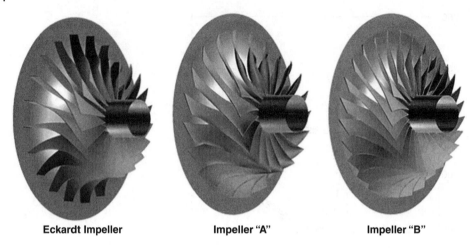

| Eckardt Impeller | Impeller "A" | Impeller "B" |

Figure 4.40 Comparison between original and optimized impellers.

4.2.4.1 Axial-Flow Turbines

Both impulse-type and reaction blading of the axial type are commonly used. In impulse-type turbine stages, gas or steam expands only in the nozzle rows, and the rotating blades are moved only due to the change of the flow direction in the blade channels. In reaction-type turbine stages, the steam expansion and speed-up of the steam flow take place in both the fixed blade and rotating blade channels approximately to an equal degree. The ratio between the available energy (enthalpy drop) for the blade row and the total available energy of the stage is termed the stage reactivity or reaction degree, R. For a purely axial impulse-type stage, $R = 0$, and for a characteristic reaction-type stage, $R = 0.5$. The optimal velocity ratio (the ratio between the circumferential rotation speed and the fictitious steam velocity corresponding to the stage's available energy) for impulse-type stages is equal to approximately 0.47. For reaction stages ($R = 0.5$), it is twice more, and hence (because the available energy is in proportion to the velocity ratio squared) the optimal enthalpy drop per stage is half that for an impulse stage of the same diameter.

Both impulse and reaction-type turbines have their inherent strengths and weaknesses. Reaction-type blading has a somewhat better internal efficiency due to lower mean velocities and a more convergent character of its steam flow. On the other hand, reaction-type stages commonly have somewhat greater steam leakages through the stage seals. In addition, the optimal available energy drop for a reaction-type stage is less than that for an impulse-type stage of the same dimensions, thus resulting in a larger number of stages compared with impulse-type turbines of the same steam conditions and output. Reaction-type turbines are also characterized by a greater axial thrust, which requires special design countermeasures.

4.2.4.2 Outflow and Inflow Turbines

In the purely radial outflow turbine, fluid enters the turbine disk axially parallel to its axis of rotation and expands radially outwards, through one stage or a series of stages mounted on the single disk. At the discharge of the last rotor the flow goes through a radial diffuser and is then conveyed through a discharge volute. From Euler's equation of the work produced by a turbine, it is clear that the radial outflow configuration has a low specific work per stage due to the reduction of the peripheral velocity while expanding the fluid.

In radial inflow configurations (Figure 4.41), work extraction is maximized by virtue of the centripetal fluid flow and the very high values of the tip speed velocities. They have some advantages over an axial turbine as they maintain a relatively high efficiency when reduced to very small sizes and can handle an elevated pressure ratio. The pressure ratio for radial turbines used in turbochargers can be up to $4:1$, but in some applications, as for example in their use in power generation systems can become $6:1$. Moreover, a radial inflow turbine is usually much cheaper than an axial-flow type. However, when it comes to wet steam radial inflow turbines suffer from tremendous limitations related to the formation of the water phase. This has limited chances to be efficiently evacuated from the machine as water tends to be centrifuged at all times: this means a tendency for the water to be directed outwards such that flow passage obstruction is likely to occur.

4.2.4.3 Axial 1D

In spite of the remarkable advances in the field of the Computational Fluid Dynamics, meanline models built upon empirical loss and deviation correlations are still one of the most reliable and effective tools to predict the performance of turbine stages with reasonable accuracy. Since computations have almost negligible computational costs, this is particularly useful in the initial design phases in order to quickly explore the search space using almost an arbitrary number of decision variables. Also, the use of meanline models gives priceless help when multi-point optimization problems are dealt with (Figure 4.41).

In meanline codes, velocity triangles and thermodynamic transformations are calculated using Euler work expressions stage-by-stage along with revised loss and deviation

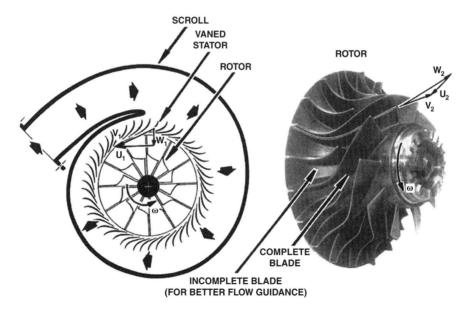

Figure 4.41 Cutaway of a radial inflow turbine. Source: From Baskharone (2006).

correlations for performance estimation (Figure 4.42). As a consequence, the preliminary turbine cross-sections, and flow path geometry are derived and assessed.

Although a detailed presentation of loss and deviation correlation is well beyond the scope of this section, a brief recall is useful to understand the basic flow phenomena and the pertinent decision variables that are typically dealt with in an optimization framework based on 1D codes.

Several loss models are available in the literature based on both algebraic models and look-up tables (derived from empirical data) to calculate turbine stage performance using a mean line approach. However, since no definitive indications about their correct usage are available, the choice of loss formulation to use in this work was made on the bases of indications given in Benini et al. (2008) and Wei (2000). Among all, the stage performance prediction model developed by Ainley and Mathieson, fully described in Ainley and Mathieson (1951), can be utilized by virtue of its simplicity and fair accuracy.

In the Ainley and Mathieson model, the total pressure loss occurring in a generic turbine blade row (in = blade row inlet, out = blade row exit) is given by (i) profile, (ii) secondary, and (iii) tip clearance losses:

$$Y = \frac{p_{in}^0 - p_{out}^0}{p_{in}^0 - p_{in}} = Y_p + Y_s + Y_{ti} \tag{4.83}$$

Profile losses are given by losses occurring at the design condition multiplied by an incidence factor whose values are given for off-design operation:

$$Y_p = \chi_i \left\{ Y_P(\alpha_{in}' = 0) + \left(\frac{\alpha_{in}'}{\alpha_{out}}\right)^2 [Y_{P(\alpha_{in}'=0)} - Y_{P(\alpha_{in}'=\alpha_{out})}] \right\} \left(\frac{s_{max}/l}{0.2}\right)^{\frac{\alpha_{in}'}{\alpha_{out}}} \tag{4.84}$$

where: $Y_{P(\alpha_{in}'=0)}$ is the profile total pressure loss occurring at zero incidence and for a blade inlet angle α_{in}' equal to 0°, α_{out} is exit flow angle, s_{max} the maximum profile thickness, l the profile chord, and α_{out} (exit flow angle) is determined from α'_{out} (blade exit angle) by means of the Carter's rule. The incidence factor χ_i is calculated by estimating the profile stalling conditions. Some correction factors are applied for shape and Reynolds effects. Secondary losses can be estimated by the following formula:

$$Y_s = \lambda \left(\frac{C_L}{t/l}\right)^2 \left(\frac{\cos^2\alpha_{out}}{\cos^3\alpha_m}\right) \tag{4.85}$$

where:

- $\lambda = (A_{out}/A_{in})/[1 + D_{hub}/(D_{hub} + 2b)]$, A_{in} and A_{out} being the projections of inlet and outlet blade channel areas in the direction orthogonal to the flow, D_{hub} the mean hub diameter and b is the mean height of the blade.

$$\alpha_{in} = \tan^{-1}\left[\frac{\tan\alpha_{in} + \tan\alpha_{out}}{2}\right].$$

- C_L is the lift coefficient of the cascade profile, which is defined as: $C_L = 2\frac{t}{l}(\tan\alpha_{in} - \tan\alpha_{out})\cos\alpha_m$, t is the blade spacing.

Tip clearance losses can be expressed as:

$$Y_{ti} = B\frac{\tau}{h}\left(\frac{C_L}{t/l}\right)^2 \frac{\cos^2\alpha_{out}}{\cos^3\alpha_m} \tag{4.86}$$

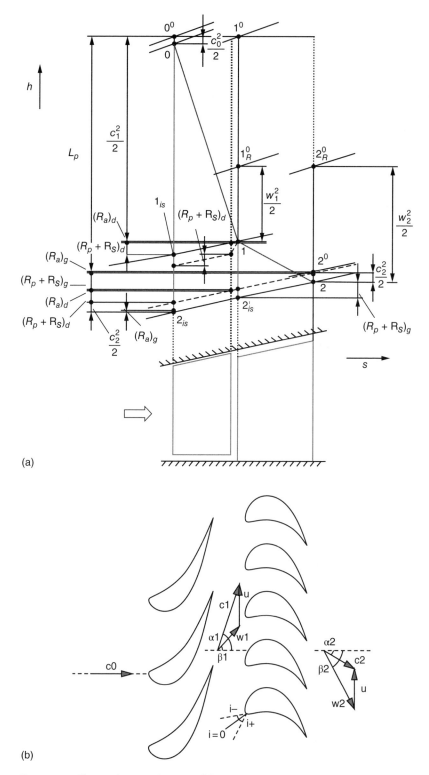

Figure 4.42 Thermodynamic diagram of the expansion in a turbine stage (a) and pertinent velocity diagram (b).

where:

- τ is the radial tip clearance for unshrouded blades.
- h is annulus height.
- B is a parameter equal to 0.25 or 0.5 for shrouded and unshrouded blades, respectively.

Further calculations are required to determine the group II losses (see Table 4.2), the influence of which will affect the ultimate amount of work available at turbine axis and, therefore, the overall stage efficiency, as better specified further on.

Once properly validated (Benini et al. 2008), a meanline model can be used to manipulate fundamental decision variables within an optimization loop in single and multi-point optimization. An example is given in Pellegrini and Benini (2013), where the influence of the following decision variables on turbine on/off efficiency and specific work has been assessed:

- Outlet stator blade angle α_{1c};
- Stator cascade solidity σ_S;
- Inlet rotor blade angle β_{1c};
- Outlet rotor blade angle β_{2c};
- Rotor cascade solidity σ_R.

Among them, cascade solidities influence mainly deviation angle, wall friction losses, and stall behavior, while blade angles influence mainly stator-rotor interaction, stall behavior, and rotor outflow conditions.

Typical objective functions that can be used along with such a model are:

Specific turbine stage work $L = h_0^0 - h_2^0$;

Total-to-static stage efficiency $\eta_{ts} = \frac{h_0^0 - h_2^0}{h_0^0 - h_{2,is}}$.

A crucial aspect of the problem is that objective functions chosen are conflicting targets; in fact, assuming u as a constant, specific work can be expressed as:

$$L = u \cdot (c_{u1} - c_{u2}) = u \cdot (w_{u1} - w_{u2}) \propto \varepsilon_R \tag{4.87}$$

However, as confirmed by the Craig and Cox loss correlations, a high deflection (ε_R) is associated with high profile losses and high secondary losses: this means that an increase in specific work causes a reduction in the efficiency. The solution of the problem will identify those values of the decision variables that give the best balance between efficiency and specific work.

Table 4.2 Turbine losses (Craig and Cox 1970).

Group I	Group II
Guide profile loss	Guide leakage loss
Runner profile loss	Balance hole loss
Guide secondary loss	Rotor tip leakage loss
Runner secondary loss	Lacing wire loss
Guide annulus loss (lap and cavity)	Wetness loss (where two-phase flow occurs)
Runner annulus loss (lap, cavity, and annulus)	Disk windage loss. Losses due to partial admission

4.2.4.4 Case Study: Multi-Point Optimization of an Axial Turbine Stage

A test case, known in literature as *E/TU-3 (AGARD AR 275)*, is chosen to assess the application of a multi-point approach in a multi-objective scenario using 1D approach. *E/TU-3* is a subsonic single-stage gas turbine whose geometrical and functional parameters are listed in Table 4.3. Note that all parameters varying with radius are given referring to their midspan values. Moreover, rotor blades are unshrouded.

The multi-point approach requires evaluating objective functions in more than one operating point (e.g., at varying of mass flow rates). A simple as well as effective approach is to use the average of each of the objective functions at different mass flow rates as global objective functions.

In this way, a bi-objective problem is formulated:

$$\max(f_{obj1}, f_{obj2}), f_{obj1} = \frac{1}{NOP} \sum_{i=1}^{NOP} L_i; \ f_{obj2} = \frac{1}{NOP} \sum_{i=1}^{NOP} \eta_{ts,ii} \tag{4.88}$$

where *NOP* is the number of operating conditions, that is, number of mass flow rates, where turbine performance is calculated.

Table 4.3 Geometrical and functional parameters for a sample turbine optimization problem.

Parameter	Value
Hub diameter – Stator [mm]	340.0
Hub diameter – Rotor [mm]	335.4
Shroud diameter – Stator [m]	450.0
Shroud diameter – Rotor [mm]	450.0
Pitch – Stator [mm]	95.5
Pitch – Rotor [mm]	62.8
Number of blades – Stator	20
Number of blades – Rotor	31
Axial distance – Stator/Rotor [m]	54.0
Inlet stator blade angle [deg]	0
Outlet stator blade angle [deg]	68.9
Inlet rotor blade angle [deg]	47.6
Outlet rotor blade angle [deg]	−57.5
Tip clearance – Stator [mm]	0
Tip clearance – Rotor [mm]	0.25

Quantity	Value
Inlet total temperature [K]	733.15
Inlet total pressure [bar]	20
Minimum corrected mass flow rate [(kg·$K^{0.5}$)/(s·bar)]	35
Maximum corrected mass flow rate [(kg·$K^{0.5}$)/(s·bar)]	47
Corrected mass flow rate step [(kg·$K^{0.5}$)/(s·bar)]	1
Rotational speed [rev/min]	7800

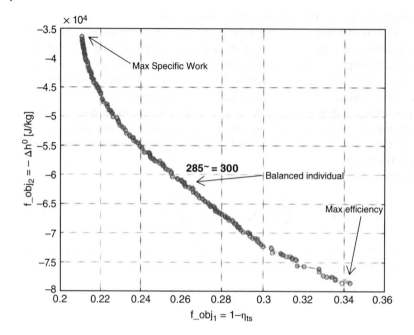

Figure 4.43 Pareto front for *E/TU-3* turbine stage multi-point optimization.

Application of population-based algorithms, such as GAs, gives an adequate solution to optimization problems. An example of a Pareto front obtained for the case study is given in Figure 4.43 after 300 generations, using 50 individuals per population.

Plots of turbine maps (Figures 4.44 and 4.45) for three individuals chosen on the last Pareto front give evidence of the validity of the approach: the first individual with maximum efficiency (see Figure 4.43), the second with a balance of efficiency and specific work, and the third with maximum specific work done.

For the three representative individuals considered here (to plot the turbine maps), corresponding decision variables values are listed in Table 4.4.

Starting from the individual with higher efficiency, moving toward the individual giving the higher specific work, one can observe that:

- α_{1c} value increases. This is reasonable as it is associated with an increase in c_{u1} and, thus, with an increase in L. At the same time, higher deflection causes reduction in efficiency.
- σ_S value increases. This is reasonable too, as variations in σ_S has similar effect to variations in α_{1c}.
- β_{1c} value increases. This is due to stator-rotor coupling: to avoid bad incidence at rotor inlet, when α_{1c} value increase, β_{1c} value must increase too.
- $|\beta_{2c}|$ value increases. This is reasonable as we know that maximum efficiency individual has almost axial absolute flow at rotor outlet, which means that increases in $|\beta_{2c}|$ value cause increases in $|c_{u2}|$ value, that means higher L, but lower efficiency.
- σ_R value increases. This can be explained in the same way of σ_S value increase.

This means that the obtained results are consistent with fluid dynamic phenomena.

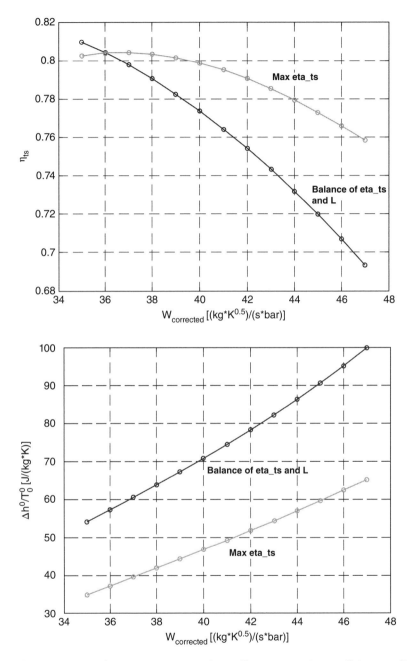

Figure 4.44 η_{ts} and L/T01 versus corrected mass flow rate – maximum efficiency and balanced individuals.

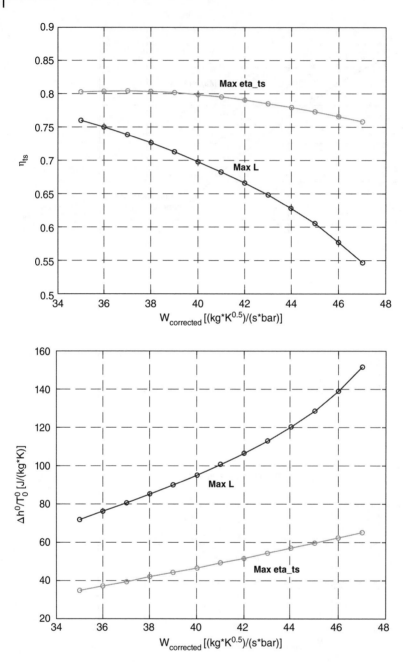

Figure 4.45 η_{ts} and L/T01 versus corrected mass flow rate – maximum efficiency and maximum specific work individuals.

Table 4.4 Decision variable values – most significant individuals.

Variable	Max η_{ts} individual	Balanced individual	Max L individual
α_{1c} [deg]	68.2	71.7	71.9
σ_S	1.12	1.41	2.45
β_{1c} [deg]	38.0	46.1	48.7
β_{2c} [deg]	−64.8	−66.8	−68.7
σ_R	1.23	1.51	2.10

4.2.4.5 Axial 2D

2D turbine cascades are divided into low-to-medium reaction types. Low reaction, or impulse, rotors are notoriously characterized by remarkable blade loadings, high-velocity flows under negligible or slightly positive pressure gradients along the main flow direction. Such phenomena negatively influence their peak efficiency, as in these bladings profile losses contribute far more severely to the stage overall efficiency compared to reaction turbines (Benini et al. 2008). As a consequence, the aerodynamic losses play an important role in determining losses of such blades and must be minimized.

To carry out such a task in an effective way, an accurate prediction of the transition zone from laminar to turbulent in blade boundary layers should be considered. Correlation-based transition models, namely the $\gamma - \theta$ laminar-turbulent model developed by Menter et al. (2006) can be used to predict transition with good approximation. According to the authors, it is suitable for low-to-medium Mach number flows, both developing around bodies in an open domain (e.g., airfoils) or within bounded walls (e.g., in turbomachinery) and delivers good performance on 2D-3D unstructured grids on both single-processor and parallel machines.

On the other hand, examples of optimization procedures and applications to reaction turbine cascades are almost uncountable in the open literature; for example, among others (Horlock 1966; Tong and Gregory 1990; Mengistu and Ghaly 2008; Oksuz et al. 2002).

Nevertheless, very few studies exist regarding design optimization of impulse turbine cascades. Chung et al. (2005) developed an optimization method for a rotor blade of a Curtis turbine where the blade section has been described only using the exit blade angle, the stagger angle, and maximum camber. The optimization procedure involved the use of a second-order response surface method and an evolution strategy for maximizing the lift to drag ratio of the blade. Lampart (2004a,b) minimized the enthalpy loss of a high-pressure (HP) steam turbine stage using a simplex method. Mengistu and Ghaly (2008) proposed an optimization method using a GA coupled with an artificial neural network for the aerodynamic improvement of a transonic impulse blading, the well-known Hobson cascade (Fottner 1990). In that paper, the flow was assumed to be inviscid and the objective function addressed the purpose of reducing the losses induced by a shock wave developing onto the blade suction side.

4.2.4.6 CFD Models: Implementation and Validation

Fluid dynamics simulations of turbine cascades are routinely carried out using RANS solvers (see e.g., ANSYS Fluent). Turbulence is treated via the previously mentioned

$\gamma - \theta$ model, which actually represents an evolution of the original k-ω SST formulation developed by Menter (1992, 1994), Menter et al. (2003) in fact, the transition model, based on local variables, relies on some empirical correlations.

Two basic transport equations, one for intermittency and one for the transition onset criteria, are used in the transitional model, the main outcome of which is the momentum-thickness Reynolds number. Basically, the method works by modifying both the production and dissipation terms of the turbulent kinetic energy typical of the SST k-ω model via a special function, known as "effective intermittency": the latter actually activates the production of turbulent kinetic energy downstream of the transition point. This allows the abrupt occurrence of transition caused by separation of the laminar boundary layer to be captured. On the other hand, the second transport equation aims to simulate the non-local influence of the turbulence intensity, which is influenced in turn by both the decay of the free-stream kinetic energy and the modifications the undisturbed flow-speed experiences outside the boundary layer region. The latter equation is used to fit the transition onset criteria to the empirical correlations in the previously mentioned intermittency equation.

Unlike the conventional turbulence models, the $\gamma - \theta$ model does not mimic the boundary layer behavior via mathematical approximations, but it rather attempts to implement some correlation-based schemes into widely used, general-purpose CFD tools. Further details on both theoretical formulation and validation of the model can be found in Menter et al. (2006).

Laminar to turbulent transition can be captured in a reliable way only by using a high-quality structured mesh close to the turbine blades' surface: specifically, the recommended values for the resulting $y+$ must not be greater than unity, otherwise the model is not guaranteed to calculate wall shear stress with a satisfactory accuracy.

Moreover, the advection scheme used for the turbulence and transition model equations strongly affects the calculated transition position: to this purpose, a bounded second-order upwind scheme is usually recommended.

Transitional model $\gamma - \theta$ can be tested first on a test case for validation purposes: one example is the Pratt and Whitney linear cascade known as "Pak B" (Huang et al. 2006). This cascade was experimentally tested in quite an extensive way: both surface pressure distributions and boundary layer velocity profiles were acquired and data were published in the open literature. The final aim of the experimental campaign was to analyze the effects of both Reynolds number and free-stream turbulence intensity on the blades' fluid dynamics behavior, with particular emphasis on separation onset and flow reattachment over the suction side. These phenomena are typical of low-pressure turbine stages blade rows operated at low Reynolds number conditions.

The Pak B blade shape is illustrated in Figure 4.46, where also the location of the static pressure probes used in the experiments is highlighted. The blade chord length is $C = 7.0$ in (17.78 cm): where the stagger angle of the blade is equal to 26°, it follows that the axial chord length is $Cx = 6.28$ in. or 15.95 cm. The cascade solidity is 1.13, while the inlet angle and the design outlet angle are equal to 55°and 30°, respectively, both measured with respect to the tangential direction.

From the experimental acquisitions, a plateau on the pressure coefficient profile over the blade suction side is clearly apparent (see circled zones in Figure 4.47), which suggests the occurrence of a separation bubble.

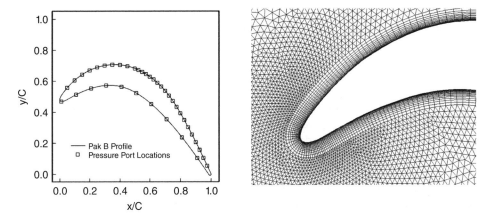

Figure 4.46 Pak B blade shape and computational mesh.

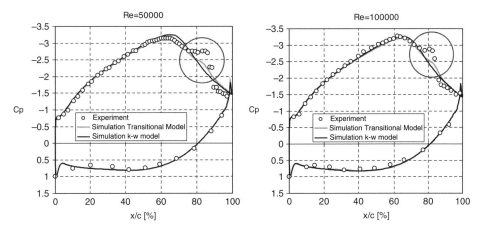

Figure 4.47 Pak B turbine: Simulated pressure coefficient distributions over Pak B blades at different Reynolds numbers and a free-stream turbulence level of 1.6%: comparison with experiment.

The "Pak B" cascade was tested at various Reynolds numbers (the Reynolds number being based on the free-stream inlet speed and blade axial chord length) and turbulence intensity levels. In the following, only the intermediate turbulence intensity, equal to 1.60%, is considered at three different Reynolds number values equal to 50 000, 75 000, and 100 000, respectively.

Two-dimensional unstructured grids are conventionally used where some high-quality structured elements rows next to the blade surface as is required in order for the boundary layer behavior to be accurately captured and the $y+$ values to comply with the previously mentioned constraints of the $\gamma - \theta$ model. A good choice is to adopt 15–20 streamwise gridlines for solving the boundary layer zone, with the first gridline some 1/100 mm distance from the wall, and imposing a grid expansion ratio of 1.2. Usually, grids with at least 20 000 elements per single passage give satisfactory results for the purposes of the optimization. A global view of a typical computational mesh is shown in Figure 4.46, together with a detail of the leading-edge region. It is worth noting that with this mesh the number of grid layers inside the physical boundary layer ranges from nine at the airfoil leading edge to 15 at the trailing edge in all the analyzed test cases. Therefore, both the velocity profiles and their derivatives inside the physical boundary layer can be accurately captured: this in turn affects the momentum-thickness Reynolds number evaluation, and consequently the transition onset location.

Validation results are summarized in Figure 4.47, where the simulated pressure coefficient profiles are compared with both experimental data and computational results coming out from a standard, fully turbulent k-ω model. It clearly appears that the transitional model captures both the separation onset location and extension over the suction side. Nevertheless, the previously mentioned plateau featured by the experimental pressure coefficient (more pronounced with decreasing Reynolds numbers) is not satisfactorily simulated. This happens for all the different turbulence intensity levels tested in the experimental campaign, and partially contradicts the results obtained by Menter and Langtry et al. (2006), where the cascade was simulated using a proprietary Navier–Stokes tool integrated with the $\gamma - \theta$ model. In fact, the separation bubble was actually well captured there at each of the Reynolds number values, especially at medium-to-high turbulence intensity levels.

4.2.4.7 Case Study: Description, Geometry Parametrization, and Meshing

A particular 2D cascade geometry used in impulse steam turbine stages is selected as a case study. The blade geometry was originally developed using multiple circular arcs and therefore seems appropriate for an optimization case. Geometric representation of this profile is an important part in the aerodynamic shape optimization procedure. The parameters in the geometric representation of the blades were used as design or decision variables in the aerodynamic optimization process.

Due to the necessity of reducing the computational time only a portion of the blade profile, namely the pressure side and the rear part of the suction side, was parameterized. Even though the validation on the suction side of the Pak B blade did not give completely satisfactory results due to not being capable of capturing the separation bubble, the choice of parameters was driven by the observation it is the rear part of the profile that is responsible for the greatest amount of total pressure losses due to the boundary layer thickening. Moreover, the parameters of the rear of the profile actually govern the dimensions of the discharge section and hence the pressure acting on the cascade back section.

Therefore, seven design variables were selected (see Figure 4.48), five on the blade pressure side, starting from the trailing edge region and moving backward until 25% of chord, and two on the suction side, once again starting from the trailing edge region. These parameters correspond to the control points of a cubic spline curve. Each parameter is given a range of variations along the local direction normal to the blade surface. To avoid the creation of blades with mass flow rates different from that of the baseline configuration, both the lower and the upper boundary of each of the design variables was set to 0.3 mm.

In Figure 4.48, a comparison among the baseline and two configurations characterized by having each of the design variables set at their upper and lower boundaries, respectively, is shown. The configuration was meshed using 16 boundary layer rows having the first layer height of 0.002 mm and a growth ratio equal to 1.3. Following these guidelines,

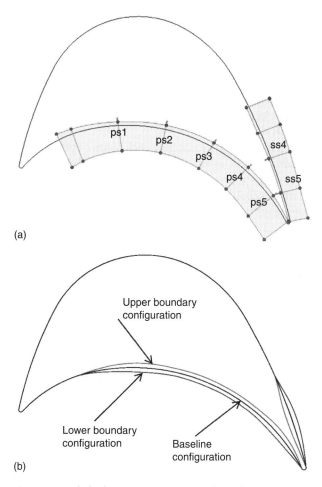

(a)

(b)

Figure 4.48 Blade shape parameterization through the morph surface technique (a). Upper and lower boundaries configurations defining the limits of the search space (b).

a 130 k-node cell grid, referred to as a "viscous" hybrid mesh (ANSYS Fluent), was built per single passage (Figure 4.46).

In the CFD simulations, the fluid is assumed as an ideal gas with constant specific heats. Similar to the validation test case, a total pressure boundary condition is applied at the domain inlet while a static pressure value is specified at the domain outlet. A no-slip condition is adopted for the blade walls, which were assumed to be adiabatic and hydraulically smooth. Iterations are performed until a second-order converged solution with a normalized residual on the continuity equation less than $1 \cdot 10^{-6}$ is obtained and the mass flow averaged air velocity at outlet reaches a stabilized value.

Updating and manipulating the mesh during the optimization is accomplished using a morphing technique, a graphic operation that can be used to transform graphical objects interpolating between two of them. In particular, a feature-based morphing technique was adopted, which entails the selection of a finite set of features in the shape of the source object, for example surfaces or volumes, and the corresponding collection on the target object. This method encloses the mesh in one or more deformable blocks (namely, the "morph surfaces"), each of which governs the movement of the mesh within its boundaries. Thus, the shape of the mesh is changed simply by modifying the shapes of the blocks, which can be attained in a very flexible way: specifically, the length and curvature of each block edge can be varied independently from the others and adjacent blocks can be linked through various continuity conditions; finally, an auto-tangent option allows the edges of adjacent blocks to morph as spline curves, therefore ensuring a smoother morphing between morph surfaces.

The morph blocks' warping is defined explicitly by mapping each feature in the source object to its correspondent in the target object, as explained in Gomes (1999) and Arad and Reisfeld (1995). Using such a technique, surface portions of the computational domain are morphed quite easily while keeping mesh distortion to a minimum. In this work, the mesh of the blade profile was actually enclosed into a series of 11 deformable 2D blocks depicted with shading in Figure 4.48.

The actual optimization scheme is operated as follows:

- First a DoE analysis (Montgomery 2012) was conducted aimed at getting as much information as possible from the least number of simulated configurations, therefore saving time, and resources. In particular, the objective of the DoE analysis were the following: (i) identifying which variables are most influential on the output, (ii) determining where to set the meaningful variables so that the output is almost near the desired normal value, (iii) finding out where to set the influential controllable factors so that the variability in output is small, and (iv) determining where to set the influential controllable factors so that the effects of the uncontrollable variables are minimized. In order to succeed in this goal, it is necessary to carry out a series of tests (or experiments) following a so-called strategy of experimentation. In this paper, both 2^k and 3^k factorial strategies, in which the k factors are varied together, were used.
- Then, a response surface methodology (RSM) was adopted to interpolate data obtained after the DoE analysis (Myers and Montgomery 1995), thereby creating a metamodel (i.e., a model of a model), often referred to as surrogate model. In this paper, a Least Squares Regression (LSR) model was built, where the objective consists of adjusting the parameters of a model function (a polynomial) so as to best fit the data set containing the simulated configurations.
- Finally, a gradient-based optimization technique, namely a sequential quadratic programming (SQP) algorithm, was used to search the optimal solution onto the surface previously created (Fletcher 1987). This approach has proved to be powerful for solving general optimization problems with smooth objective and both constrained and unconstrained functions. Of course, this is classified as a local optimization technique; however, when used in combination with a DoE/RSM method, it turns out to be very effective, since it fast explores only the most promising regions, where the minimum is likely to be positioned. Using such a procedure actually saves computational time because the SQP algorithm does not require too much effort to evaluate the Hessian of the Lagrangian matrices of the objective function with respect to the decision variables.

4.2.4.8 Results

The DoE analysis was performed in two steps, described in the following subsections.

4.2.4.8.1 Preliminary DoE Analysis

First, for each of the seven variables involved in the shape parameterization, only two levels were investigated. These were the lower boundary values, corresponding to a deformation of −0.3 mm over the baseline configuration, and the upper boundary ones, corresponding to a deformation of +0.3 mm. Hence, the total number of combinations investigated was $2^7 = 128$ cases. From the preliminary screening, the most influential factors are determined using a statistical analysis based on the Student t parameter:

$$t = \frac{abs(\bar{x}_1 - \bar{x}_2)}{\sigma} \tag{4.89}$$

where $\bar{x}_1$ is the mean value of the upper set, namely the set of the configurations where the investigated variable is given the upper value (+0.3 mm in this case); $\bar{x}_2$ is the mean value of the lower set (corresponding to −0.3 mm in this case), and σ is the standard deviation, defined as follows:

$$\sigma = \sqrt{\frac{\left(\sum_{i=1}^{n_1}(x_{1i} - \bar{x}_1)^2 + \sum_{i=1}^{n_2}(x_{2i} - \bar{x}_2)^2\right)(n_1 + n_2)}{(n_1 + n_2 - 2)n_1 n_2}} \tag{4.90}$$

As is well-known, the Student t parameter is defined in such a way that the bigger its value, the higher the response variation caused by the corresponding parameter. In Figure 4.49, the Student t values are presented for each of the investigated design variables. DoE analysis helps to point out that the influence of the variables ps1 and ss4 is negligible when compared to the other ones. It is up to the designer to decide whether to keep the latter parameter as a design variable in the performed analyses, since otherwise only one parameter would be left to control the suction-side shape.

It clearly appears that the design variable ps1 gives no significant contribution to the fitness function value. Hence, it was definitively discarded in the following analyses, in particular, in the second DoE analysis, which was carried out to explore the design space in more depth.

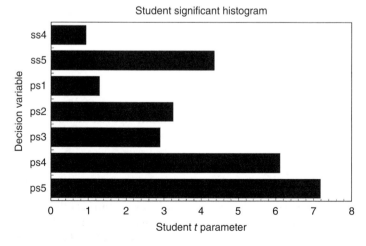

Figure 4.49 Student histogram based on first DoE analysis.

4.2.4.8.2 Second DoE Analysis

This was a full factorial DoE study extended to all the parameters except ps1 and involving three levels per each variable (the upper limit, the lower limit, and the intermediate level corresponding to the baseline shape). Therefore, the total number of cases investigated amounted to $3^6 = 729$. The usefulness of the Student analysis in reducing the overall optimization procedure time becomes evident, since the number of cases to be investigated if all the initial parameters were accounted for would have been three times larger. As a matter of fact, a huge amount of computational resources was saved by skipping over the parameter ps1 in this analysis.

4.2.4.9 RSM

Starting from the second DoE analysis, an approximation model was created in order to build a response surface using the least square regression criterion. Such a model was used to perform the optimization study by using far less computational resources than an optimization based on a direct solver call. A least-square-regression-based surrogate model in the of order p may be written as:

$$f = A\beta \tag{4.91}$$

While the response vector at the n sampling points is:

$$y = A\beta + \varepsilon \tag{4.92}$$

where y is a $(n \times 1)$ vector of output responses obtained from the DoE, A is an $(n \times p)$ matrix obtained from input values of the DoE, β is a $(p \times 1)$ vector of regression coefficients and ε is a $(n \times 1)$ vector of random errors. The least squares estimator of β is obtained by weighted least squares fitting of the response surface f into the set of responses y at the sampling points:

$$(y - f)'W(y - f) \Rightarrow \min \tag{4.93}$$

This is equivalent to solving the system of normal equations:

$$\beta = (A'WA)^{-1}A'Wy \tag{4.94}$$

where W is a diagonal $(n \times n)$ matrix of weight coefficients. In the conventional LSR, all weights are set to unity so that the system of normal equations finally becomes

$$\beta = (A'A)^{-1}A'y \tag{4.95}$$

4.2.4.10 SQP

SQP methods are the state of the art in nonlinear programming techniques for constrained optimization problems. Starting from a quadratic approximation of the Lagrangian function, SQP identifies a search direction s by solving the corresponding quadratic sub-problem:

$$\min f(s) = f(x_k) + \nabla_{x}^{T}f \cdot s + \frac{1}{2}s^T C_{(k)}s \tag{4.96}$$

where $C_{(k)}$ is a positive definite matrix, which is equal to the identity matrix at the beginning of the search iterations and is successively updated (Powell 1978) during the optimization loop until it becomes a reliable approximation of the Hessian of the Lagrangian function.

Once the new search direction s is found, this is used to define a new k-th solution from the previous one:

$$x_k = x_{k-1} + \alpha s \tag{4.97}$$

The step length parameter α is determined through an appropriate line search procedure aimed at achieving the maximum possible decrease in the merit function.

Specifically, the algorithm works as follows:

1. Initialize $C = I$.
2. Compute all gradients.
3. Solve the quadratic programming sub-problem.
4. Compute the Lagrange multipliers.
5. Check for convergence.

The results obtained from the application of the SQP algorithm are reported in Table 4.5, where the optimization history is summarized in terms of the design variable values and the corresponding fitness function score, while in Figure 4.50 the convergence history of the loss coefficient is given.

Design variable values are normalized in such a way that 1 corresponds to the upper bound (namely +0.3 mm), −1 to the lower bound (namely −0.3 mm), while 0 stands for the non-deformed configuration.

As evidenced in Table 4.5, the relatively simple optimization procedure which was applied, based on the LSR approximation model, allowed to gain a loss reduction up to 16%.

In order for these promising results to be verified, a direct CFD simulation of the optimized configuration (namely the one corresponding to the 12th iteration) was performed, which confirmed this trend.

Table 4.5 Optimization history table.

Iteration	ps5	ps4	ps3	ps2	ss5	ss4	Ainley loss coefficient
1	0.0000	0.0000	0.0000	0.0000	0.0000	0.0000	0.0893
2	0.1007	0.0224	0.0448	0.0504	−0.0504	−0.0112	0.0879
3	0.6148	0.0704	0.1408	0.1584	−0.1584	−0.0352	0.0840
4	0.9109	0.0919	0.1838	0.2068	−0.3073	−0.0459	0.0832
5	1.0000	0.1410	0.2127	0.2441	−0.3299	−0.0474	0.0829
6	1.0000	0.2923	0.2571	0.3145	−0.3023	−0.0722	0.0820
7	1.0000	1.0000	0.4798	0.6484	−0.1928	−0.2059	0.0798
8	1.0000	1.0000	0.5026	0.6592	−0.2083	−0.2086	0.0797
9	1.0000	1.0000	0.7406	0.7675	−0.3595	−0.2584	0.0790
10	0.9692	1.0000	0.8656	0.8293	−0.4139	−0.2895	0.0789
11	0.9643	1.0000	0.9015	0.8545	−0.4152	−0.3066	0.0789
12	0.9677	1.0000	0.9564	0.9089	−0.3950	−0.3468	0.0789

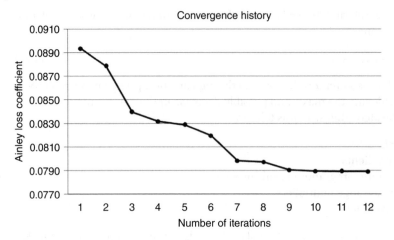

Figure 4.50 Convergence history using SQP.

The results obtained are reported in Table 3, along with the Best-DoE configuration (namely the 721st configuration) and the baseline configuration.

A reduction in the losses of nearly 13% was achieved with both the Best_DoE and the optimized configuration Best_SQP (LSR-based, recalculated by means of CFD code). This can be explained by the fact that the Best_SQP is a direct optimization strategy driven by the CFD results, and thus the minimum of the objective function is found in quite a reliable way based on the fluid dynamics computations; on the other hand, the comparable or even apparently better results an RSM may give can be due to the function minimum being estimated in a more approximate way, which is intrinsic to the method. The latter issue is relevant, since nearly the same results were achieved with a slightly different blade configuration. This is clearly shown in Figure 4.51, which displays the three different blade geometries.

Even though the Best_DoE configuration provided about the same results as the SQP optimized one, its shape is not as streamlined as the optimized one; hence its aerodynamic behavior should be investigated further.

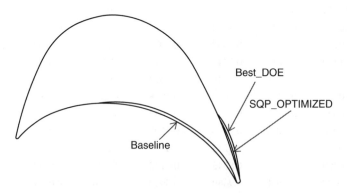

Figure 4.51 Comparison of three different blade profiles.

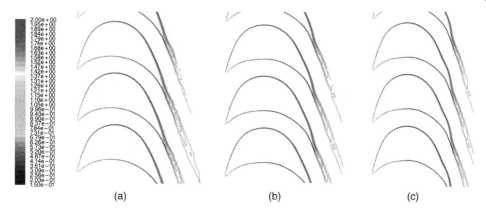

Figure 4.52 Loss coefficient: (a) Baseline (b) Best_SQP (c) Best_DOE.

In order to compare the different flow-fields characterizing the three previous configurations, the Ainley loss coefficient Y was examined, as reported in Figure 4.52. From the contour plots, a clear reduction in the total pressure losses in the blade wake, which is significantly less thick, was obtained in both the Best_DoE and the Best_SQP. This is attributed to a more favorable pressure gradient toward the rear of the suction side.

It should be observed that, due to the coarse resolution of the DoE points, variations in the loss coefficient were not expected to be accurately captured for the extremely small geometric changes involved. On the other hand, when using surrogate models, a compromise between accuracy and computational efficiency commonly needs to be reached. In this context, despite the interpolated loss coefficient (see Table 4.6, Best_SQP [LSR-based]) was quite different from the recalculated one (see Table 4.6, Best_SQP [LSR-based] CFD recalculated), the approximated model can still be considered a suitable way for searching the minimum of the objective function, while requiring far less computational resources than a direct CFD-based optimization. Moreover, the optimal parameter settings identified in this phase could be used as a starting point for performing a direct optimization: this approach can result in fewer iterations for convergence to be reached, leading to substantial saving of computational resources.

Table 4.6 Resume table.

Configuration	Design variables						Ainley loss coefficient	Reduction [%]
	ps5	ps4	ps3	ps2	ss5	ss4		
Baseline	0	0	0	0	0	0	0.0940	–
Best_DOE	1	1	1	1	−1	−1	0.0819	12.88
Best_SQP (LSR-based)	0.9677	1	0.9564	0.9089	−0.395	−0.347	0.0789	16.10
Best_SQP (LSR-based) CFD recalculated	0.9677	1	0.9564	0.9089	−0.395	−0.347	0.0820	12.83

4.3 Fans

Fans and compressors are the machines that convert external mechanical energy to static and/or dynamic energy of a gas. They are classified as the machines with a lower pressure rise (less than 100 kPa) compared to compressors. Sometimes, fans with a pressure rise in a range of 10–100 kPa are called "blowers." The main function of fans is to transport a required amount of gas to a specified location using lift and/or centrifugal force by rotating fan blades. Therefore, fans are used in a variety of applications; for transport working fluids in industry, for ventilation in buildings, tunnels, and cars, and for cooling in electric and electronic devices. Since fans are distributed widely everywhere, optimization of fan design contributes greatly to the reduction of worldwide electric energy consumption.

4.3.1 Centrifugal, Axial-Flow, Mixed-Flow, and Cross-Flow Fans

Fans are usually classified into axial-flow, centrifugal, and cross-flow types according to their structure. Among these types, the axial-flow, and centrifugal types are the most widely used fan types. The most popular type of fan used up to the Second World War was the axial flow, which had efficiencies of over 80%; for example, only axial-flow fans were used for induced draft (Eck 1973). The old design of centrifugal fans was not acceptable in terms of the dimensions and process compared to axial-flow fans. However, a series of design innovations performed thereafter, such as implementing the design of centrifugal pumps and employing airfoil blades, made centrifugal fans compact and enhanced their efficiency up to 90%. Due to this remarkable development in their design, currently, centrifugal fans form the majority.

4.3.1.1 Axial-Flow Fans

An axial-flow fan (or simply an axial fan) is defined as a fan where both incoming and outgoing flows from it are substantially parallel to the axis of the fan. Axial-flow fans are appropriate for applications that require large flow rates at relatively low-pressure rise compared to centrifugal fans. The advantage of axial-flow fans is high efficiency, while disadvantages are relatively high noise and rapid decrease in the efficiency at off-design conditions. Figure 4.53 shows two different types of ducted axial-flow fan; vain-axial and tube-axial fans depending on the location of the motor. In a vane-axial fan, stators are installed just downstream of the rotors to remove the swirling flow in order to enhance efficiency and pressure rise. These stators can also be added upstream of the rotors.

Typical performance curves of a fan are presented schematically in Figure 4.54. The design point of a fan coincides with the point of the maximum efficiency. If the fan is operating at a flow rate smaller than that of maximum pressure point, there is a danger of approaching the stall point. In the surge region (left of stall point), complex flow structures with massive flow separations in the fan occur that cause serious vibrations.

The propeller fan is also a kind of axial-flow fan, which operates without a duct. Thus, these fans have most simple structure of all. Small propeller fans are used for simple air circulation/ventilation and cooling in electronic devices, electric motors, cars, and so on; large propeller fans are used in cooling towers. The efficiency range of these fans is from 60% for sheet metal types to 75% for the larger units (Osborne 1973).

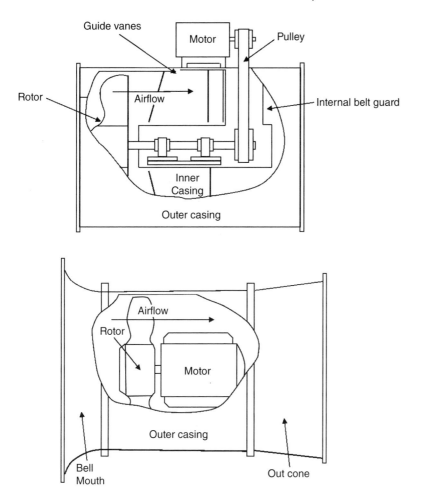

Figure 4.53 Two different types of ducted axial-flow fan (vain-axial and tube-axial fans).

4.3.1.2 Centrifugal Fans

In a centrifugal fan, the air enters the fan in the axial direction but is discharged in the circumferential direction. Centrifugal fans have an advantage of HP rise but with small flow rate compared to axial-flow fans. Centrifugal fans are classified as forward-curved, backward-curved, and radial blade depending on the direction of blade curvature in the impeller as shown in Figure 4.55.

In the forward-curved blades fans, so-called sirocco fans, which have the blades inclined toward the rotational direction in the impeller, there are a number of wide and short chord length blades (a number in the range of 30–60) in the impeller due to high loading on a single blade. These fans have relatively low noise and are of compact size, and have the particular advantages of the largest pressure rise and the largest flow rate among centrifugal fans of the same impeller diameter and rotational speed.

On the other hand, the range of blade number in backward-curved and radial blade fans is reduced greatly (to a range of 6–16) with reduced width and an increased chord length of blades, compared to those of forward-curved blade fans due to reduced blade

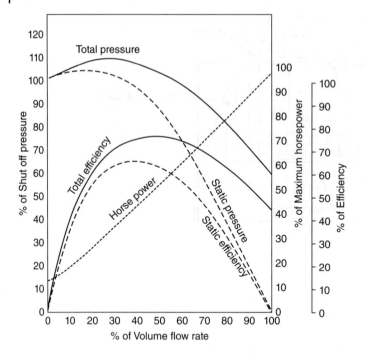

Figure 4.54 Typical performance curves of an axial-flow fan (SP: static pressure, TP: total pressure, BHP: brake horsepower, SE: static efficiency, ME: mechanical efficiency).

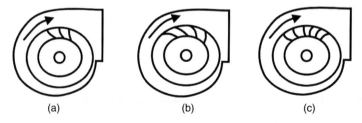

Figure 4.55 Centrifugal fan types. (a) Radial blade, (b) backward-curved, and (c) forward-curved blade centrifugal fans.

loading. In spite of the increase in fan size, backward-curved blade fans have the highest efficiency among centrifugal fan types. To further enhance the efficiency and reduce the noise, airfoil shaped blades are used in the backward-curved blade fans at increased manufacturing cost. The radial blade fans use flat blades in the radial direction. The size and efficiency of these fans fall between those of forward-curved and backward-curved blade fans. Figure 4.56 shows performance curves of different centrifugal fans.

4.3.1.3 Mixed-Flow Fans

Fans where the flow direction changes by less than 90° through the impeller are called mixed-flow fans. These are known to have pressure rise and efficiencies of values no larger than those of centrifugal fans and no less than those of axial-flow fans.

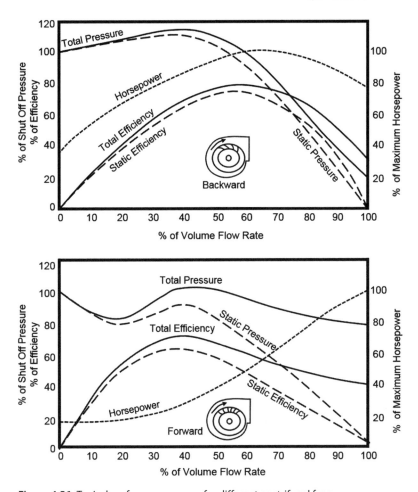

Figure 4.56 Typical performance curves for different centrifugal fans.

4.3.1.4 Cross-Flow Fans

In a cross-flow fan, the flow moves across the impeller transversely as shown in Figure 4.57. This type was patented by Mortier (1893). The cross-flow fan has been widely used in HVAC devices due to high flow rate, low noise, and compact size, although the efficiency and pressure rise is low. The impeller in this fan is usually long enough and the flow through the impeller can be considered 2D. The flow rate can be easily controlled by changing the length of the impeller.

4.3.2 Fan Pressure, Efficiency, and Laws

Performance of a fan is represented by efficiency, fan pressure, gas power, and so on. Fan pressure is not a ratio of absolute pressure at the fan outlet to fan inlet, but a pressure rise through the fan. There are three fan pressures, which are defined as follows;

- *Fan total pressure (p_t).* Difference between the total pressures at fan inlet and fan outlet.

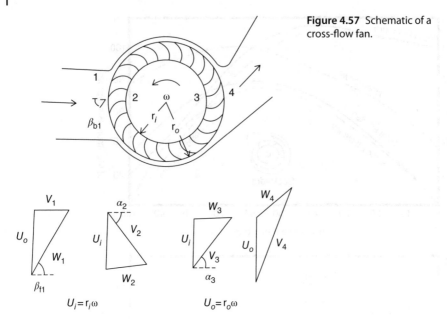

Figure 4.57 Schematic of a cross-flow fan.

- *Fan dynamic pressure.* $\tfrac{1}{2}\rho V^2$ where V is average velocity at fan outlet.
- *Fan static pressure* (p_s). The difference between fan total pressure and fan dynamic pressure (not difference between static pressures at fan inlet and fan outlet).

The air power is defined as $P = \rho Q$, where ρ is gas density and Q is volumetric flow rate. The total power of a fan is calculated by using the fan total pressure and called total air power, while it is called static air power if static pressure is used for the pressure.

Fan efficiency is defined as a ratio of the gas power (output) to mechanical power input, that is, shaft power, as follows;

- Total efficiency: $\eta = p_t Q/\text{shaft power}$
- Static efficiency: $\eta_s = p_s Q/\text{shaft power}$

In performance tests on fans, it is not easy to test for all the possible rotational speeds and all the possible inlet densities. However, fan performances at the rotational speeds and densities that are different from those of actual performance tests can be predicted quite accurately by using so-called fan laws:

$$
\begin{aligned}
\frac{Q_p}{Q_m} &= \frac{N_p}{N_m}\left(\frac{D_p}{D_m}\right)^3 \\
\frac{P_p}{P_m} &= \left(\frac{N_p}{N_m}\right)^2\left(\frac{D_p}{D_m}\right)^2\frac{\rho_p}{\rho_m} \\
\frac{P_p}{P_m} &= \left(\frac{N_p}{N_m}\right)^3\left(\frac{D_p}{D_m}\right)^5\frac{\rho_p}{\rho_m}
\end{aligned}
\tag{4.98}
$$

where the subscripts p and m indicate two different operating conditions, and D is the diameter of the impeller.

4.3.3 Aerodynamic Analysis of Fans

There have been numerous aerodynamic investigations of various fans using both numerical and experimental methods so far. In particular, with the development of CFD and computers, 3D RANS analysis became popular in the analysis of fan flows partly replacing experimental tests in the last two decades.

4.3.3.1 Axial-Flow Fans

CFD using 3D RANS analysis could be more easily applied to the analysis of axial fans due to the relatively simple flow structure through the fan blades compared to other fan types. However, other simplified analysis methods considering the flow characteristics of various axial fans have also been developed to reduce computing time.

Sorensen and Sorensen (2000) developed an efficient numerical model for the aerodynamic analysis of rotor-only axial fans. The model was based on a blade-element principle whereby the rotor is divided into a number of annular stream tubes. In this model, relations among flow properties, such as pressure, velocity, and radial position, are derived for each stream tube from the conservation equations for mass, tangential momentum, and energy. In the blade-element principle, the governing equations are only solved at one axial station using blade forces obtained from tabulated cascade-airfoil characteristics. They suggested that calculation with this model was at least 100–200 times faster than the SLM suggested by Novak (1967), which requires the solutions at multiple axial stations. Numerical results with this model agreed well with experimental data for efficiency and total pressure rise as shown in Figure 4.58. However, some discrepancies were found near the hub and blade tip as shown in Figure 4.59, due to the inviscid nature of the model. The blade-element principle was used widely for the design of low-pressure axial-flow fans. The so-called free-vertex method (Wallis 1961) was suggested for fan designs that have no radial flow, but this method is only successful when the fan actually satisfies free vortex requirements. For other fans with arbitrary vortical flows, Wallis (1961) suggested an analysis method with approximately linear radial distributions of axial and tangential velocities. Downie et al. (1993) also proposed a model for analysis of a single rotor fan.

Figure 4.58 Measured and calculated (based on a blade-element principle) efficiencies at various flow rates in a rotor-only axial fan. Source: Reproduced with permission from Sorensen and Sorensen (2000). Copyright © 2000 by American Society of Mechanical Engineers.

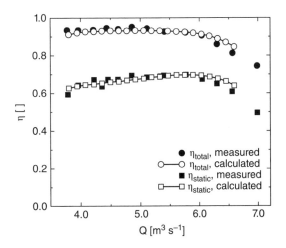

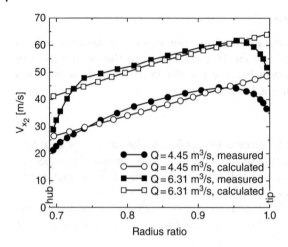

Figure 4.59 Spanwise distributions of axial velocity far downstream of the rotor at two flow rates in a rotor-only axial fan. Source: Reproduced with permission from Sorensen and Sorensen (2000). Copyright © 2000 by American Society of Mechanical Engineers.

Estevadeordal et al. (2000) studied experimentally the flow characteristics in a low-speed axial fan using digital particle image velocimetry (DPIV). The blade position in PIV method was synchronized by using the laser pulses and a charge-coupled device (CCD) camera shutter for examining instantaneous and time-averaged velocity quantities. They concluded that the DPIV measurement precisely captured the viscous effects, such as trailing edge wake structures and the suction and pressure-side separation.

Meyer and Kröger (2001) performed a numerical and experimental investigation of the flow field in an axial fan. The numerical calculations were carried out using 3D RANS analysis with a k-ε turbulence model in order to find the effects of stagger angle of the blade on the performance curve of the axial fan. In general, the numerical results showed good agreement with the experimental data, but the fan power consumption determined by numerical calculation was slightly lower than experimental data due to the additional tangential force associated with blade tip clearance as shown in Figure 4.60.

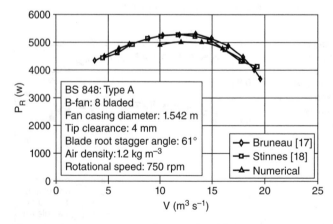

Figure 4.60 Fan power versus flow rate for a blade root stagger angle of 61° in an axial fan. Source: Reprinted from Meyer and Kröger (2001) (figure 16 from original source), © 2001, with the permission of John Wiley & Sons, Inc.

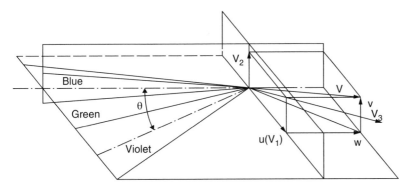

Figure 4.61 Velocity-component definitions for PDA. Source: Reproduced with permission from Zhu et al. (2005). Copyright © 2005 by American Society of Mechanical Engineers.

Zhu et al. (2005) experimentally and numerically investigated the flow field; in particular, the tip leakage flow of an axial ventilation fan. As an experimental method, a 3D phase Doppler anemometer (PDA) system (Wisler and Mossey 1973; Murthy and Lakshminarayana 1986; Stauter 1993), which consists of a laser source, transmitting and receiving optics, and a signal processor, was adopted to observe the flow field inside the tested fan. Three different wavelengths of beam (green, blue, and violet) were arrayed to calculate the axial (u), tangential (v), and radial (w) components of the velocity from the following formulation (as shown in Figure 4.61);

$$u = V_1$$
$$v = V_2$$
$$w = V_3/\sin\theta - V_1/\tan\theta \tag{4.99}$$

All laser anemometer results were transferred to the velocity components, and the average velocity distribution was obtained by an ensemble-averaged measurement for each individual blade passage. Three-dimensional RANS analysis with k-ε turbulence model was performed using a commercial CFD software package, CFX-TASCFLOW 2.09, which uses the finite volume method and algebraic multigrid method based on the additive correction multigrid strategy. The numerical results for the performance curve and the structure of tip leakage flow showed good agreements with experimental data. They found that the static pressure and efficiency degenerated with an increase of tip clearance of the tested fan, and the tip clearance led to the large leakage vortex that caused flow losses and blockage to the main flow near the tip region, as shown in Figure 4.62.

Jang et al. (2005) investigated the flow characteristics of tip leakage and wake flows in a low-speed axial fan using an experimental analysis. A hot-wire sensor consisting of a tungsten filament wire of 5 μm with I and L type supporters, indicated in Figure 4.63, was used to obtain the mean velocity and fluctuation inside and downstream of the fan rotor. A five-hole probe was set up to measure 3D velocity distributions upstream and downstream of the fan rotor and was calibrated over a range of pitch angle. As the results of the experimental worked, it was confirmed that significant axial velocity degeneration near rotor tip region was observed at near stall condition due to a large blockage effect. Also, it was found that the trajectory of tip leakage vortex was significantly affected by

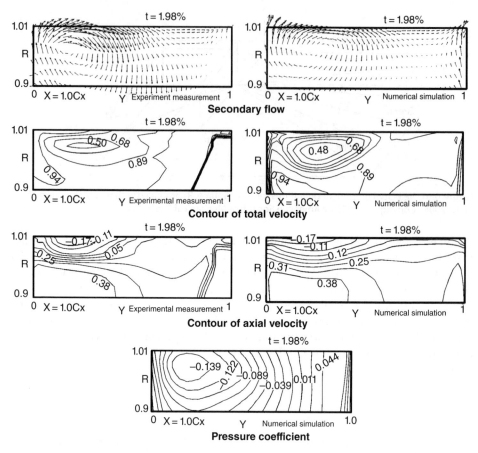

Figure 4.62 Secondary flow, contour of axial velocity, and total velocity on the surface at 100% axial chord location after the leading edge of an axial-fan blade for the normalized tip clearance of 1.98%. Source: Reproduced with permission from Zhu et al. (2005). Copyright © 2005 by American Society of Mechanical Engineers.

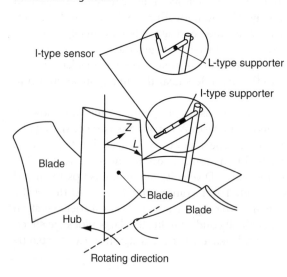

Figure 4.63 Test blade measuring system. Source: Reproduced with permission from Jang et al. (2005). Copyright © 2005 by American Society of Mechanical Engineers.

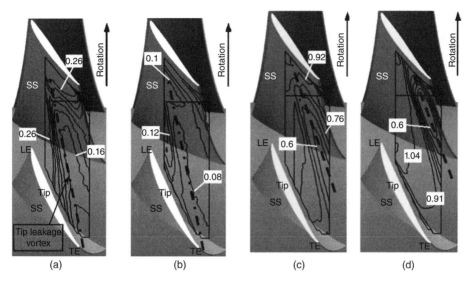

Figure 4.64 Contours of velocity fluctuation on the plane 96% span in a low-speed axial fan. (a) $\Phi = 0.28$, (b) $\Phi = 0.31$, (c) $\Phi = 0.41$, and (d) $\Phi = 0.47$. Source: Reproduced with permission from Jang et al. (2005). Copyright © 2005 by American Society of Mechanical Engineers.

the flow rate, and the tip leakage vortex led to a disruption of formation of wake flow at a low flow rate and high-velocity fluctuation near the suction surface at a high flow rate, as shown in Figure 4.64.

Corsini and Rispoli (2005) performed a numerical analysis of a HP axial ventilation fan using 3D RANS analysis. To test turbulence closure models, the performance of a non-isotropic closure model (called CLS96 model in this study), was compared with that of the classical linear k-ε model (LS74 model). Validation of numerical results using experimental data confirmed that CLS96 model predicted the boundary layer emerging from the linear prediction well compared to LS 74 model as shown in Figure 4.65. The results also showed that CLS96 model was more accurate than LS74 model in predicting the swirl flow at the design and low flow rates. It was suggested in this work that the nonlinear model could improve the prediction of stall-like behavior on the blade suction side as well as the leakage phenomena.

The effects of blade sweep on the fan performance were investigated in many types of research (Choi et al. 1997; Envia and Kerschen 1986; Kouidri et al. 2005; Wright and Simmons 1990). Choi et al. (1997) confirmed that blade sweep of a propeller fan was effective in enhancing the aerodynamic performance and reducing aeroacoustic noise through an optimization of the blade shape. Kergourlay et al. (2006) also studied experimentally the effects of blade sweep on aerodynamic and aeroacoustic performances of a propeller fan. A spectral analysis was also carried out to investigate the turbulent structure. In this work, three types of blade, which have a radial sweep, forward sweep, and backward sweep, as shown in Figure 4.66. A 2D hot fiber film probe was installed to measure velocity components with a sampling frequency of 25 kHz, which allowed 750 samples per revolution. They showed that the sweep had a great influence on both the aerodynamic and aeroacoustic performances of the fan. The turbulent kinetic energy downstream of the fans shown in Figure 4.67 shows that the turbulent kinetic energy

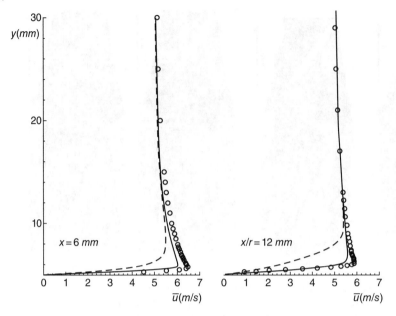

Figure 4.65 Predictions for streamwise mean velocity profiles at $x = 6$, 12 mm in a HP axial fan. (symbols: experiments; dashed lines: LS74; solid lines; CLS96). Source: Reprinted from Corsini and Rispoli (2005), with permission from Elsevier.

Figure 4.66 Fan models with different blade sweeps (backward, radial, and forward sweep). Source: Reprinted from Kergourlay et al. (2006), with permission from Elsevier.

of the forward sweep fan was generally lower than the other two cases. For the same fan, Hurault et al. (2010) performed a numerical study for examining the flow structure near the rotor in detail. For numerical analysis, 3D RANS analysis was performed using a commercial CFD code, FLUENT 6.3. The Reynolds stress model was used as a turbulence closure model to investigate the effects of anisotropic Reynolds stresses on the flow field. It was confirmed that the Reynolds stress model was more accurate in predicting turbulence kinetic energy, which was used as input data for a noise prediction (Fedala et al. 2006), than the k-ω turbulence model.

On the other hand, the advantage of blade skew for turbomachinery was also proved by many researchers (Cai and Xu 2001; Cai et al. 2003; Yang et al. 2007b, 2008).

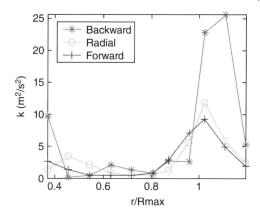

Figure 4.67 Turbulent kinetic energy (per unit mass) downstream of the fans. Source: Reprinted from Kergourlay et al. (2006), with permission from Elsevier.

Yang et al. (2007b) studied the aerodynamic performance and exit flow field of a low-pressure axial fan with circumferentially skewed rotor blades using experimental analysis. Three different rotors, that is the forward-skewed, backward-skewed, and radial rotors, were evaluated to find the effects of blade skew on the aerodynamic performance. They showed that the forward-skewed blade could delay the onset of stall, while the backward-skewed blade could accelerate the onset of stall. Yang et al. (2008) investigated the forward-skewed blade in a low-pressure axial fan using both numerical and experimental analyses. An experiment was carried out for aerodynamic and aeroacoustic analyses in an anechoic chamber. And, a numerical analysis was also performed based on 3D RANS analysis with a Spalart–Allmaras turbulence model (Spalart and Allmaras 1994). The results showed that the forward skew of the blades in the axial fan increased fan efficiency and total pressure by 1.27% and 3.56%, respectively. Furthermore, the stable operating range was also significantly extended by more than 30% and the aerodynamic noise was reduced by 6 dB(A) compared to the radial fan.

Oro et al. (2007) performed an experimental analysis of the interaction between fixed and rotating blade rows in an axial fan with inlet guide vanes. Measurements of axial and tangential velocity components were obtained by hot-wire anemometry at two different locations: one between the guide vanes and rotor blades, and the other downstream of the rotor blades (Figure 4.68). Analysis of the measured data provided a description of the mechanisms related to stator-rotor interaction and wake transport phenomena as shown in Figure 4.69.

Jang et al. (2008) carried out a performance analysis of an axial fan using 3D RANS analysis. Two hub-caps of right-angled and rounded front shapes were tested to find the effects of distorted inlet flow induced by the cap on the aerodynamic performance of the fan. The analysis results indicated that a large recirculating flow caused by the right-angled hub shape upstream of the fan rotor induced flow separation on the rotor blades near the hub, and this deteriorated the fan performance. The characteristics of the inflow depended on the distance between hubcap and leading edge of rotor blades.

Liu et al. (2010b) developed a computational approach called the "downstream flow resistance" (DFR) method to enhance the accuracy of prediction for aerodynamic performance of an axial fan using a commercial CFD code STAR-CD. The AMCA Standard 210-99 (AMCA 1999) indicates that multiple nozzles, which measure the flow rate downstream of the rotor, induce the aerodynamic losses, and adequate corrections for

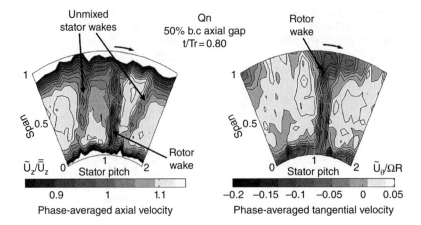

Figure 4.68 Unsteady flow patterns in axial and tangential velocity distributions at rotor exit in an axial fan with IGVs. Source: Reproduced with permission from Oro et al. (2007). Copyright © 2007 by American Society of Mechanical Engineers.

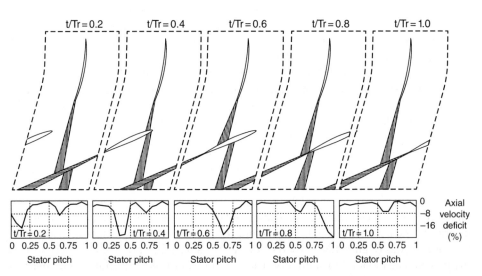

Figure 4.69 Sketch of the transport and convection of an IGV wake through a rotor passage in an axial fan. Source: Reproduced with permission from Oro et al. (2007). Copyright © 2007 by American Society of Mechanical Engineers.

these aerodynamic losses are needed. In this method, the source terms of the momentum equations are corrected in each iteration to consider the actual aerodynamic losses caused by the actual experimental test rig. Note that the method that adopted the flow resistance as a source term was a common technique, but the concept that used it for fan performance prediction was a new approach. In the DFR method, the accuracy for predicting the performance curve was significantly improved compared to the conventional method, as shown in Figures 4.70 and 4.71.

Sarraf et al. (2011) investigated experimentally the effects of blade thickness on the aerodynamic performance of an axial propeller fan. Based on the ISO 5801 standard

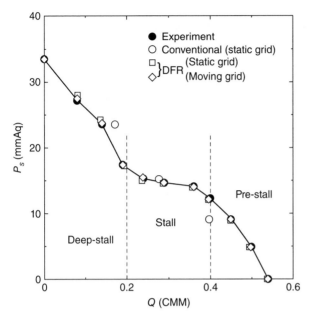

Figure 4.70 Calculated and measured fan performance curves of an axial fan. Source: Reprinted from Liu et al. (2010a), with permission from Elsevier.

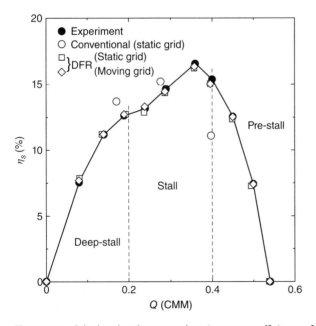

Figure 4.71 Calculated and measured static pressure efficiency of an axial fan. Source: Reprinted from Liu et al. (2012a), with permission from Elsevier.

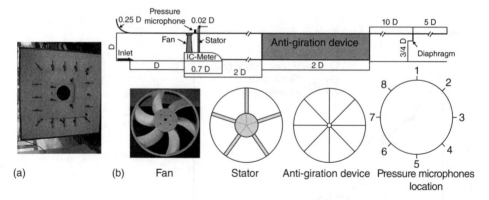

Figure 4.72 ISO-5801 test benches of an axial propeller fan. (a) Open-flow facility. The dimensions are 1.3 × 1.3 × 1.8 m. The fans suck the flow through the test bench. (b) Ducted-flow facility. This configuration is dedicated to the local measurements of wall pressure fluctuations. The fans blow the flow into the pipe. Source: Reprinted from Sarraf et al. (2011), with permission from Elsevier.

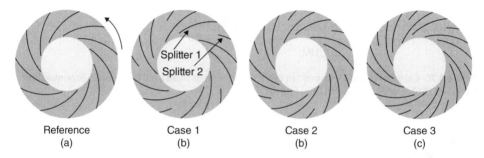

Figure 4.73 Front views of centrifugal impellers having different numbers of the main blades and splitter blades (a) Reference, (b) Case 1, (c) Case 2, and (d) Case 3. Source: Reproduced with permission from Kim et al. (2012a), IJFMS.

(AFNOR 1999), experimental apparatus was constructed to determine the overall performance of the axial fan. Laser Doppler Anemometry (LDA) using DANTEC "Flow Explore" system was adopted to measure the mean velocity field in the wake region. Wall pressure fluctuations for spectral analysis were measured by eight microphones, which were mounted downstream of the rotor, as shown in Figure 4.72. From the overall performance test, it was confirmed that increase in blade thickness resulted in an increase in the pressure rise at a part-load flow rate, which is beneficial for delaying stall onset. However, thick blades induced large tangential velocity amplitude downstream of the rotor, which caused strong peaks at blade passing frequency and its harmonics.

4.3.3.2 Centrifugal Fans

Due to the advantage of HP rise of centrifugal fans, many types of research have been performed for the analysis and design of centrifugal fans. Some of them focused their research on rotating stall in centrifugal fans. Kubo and Murata (1976) used a hot-wire probe to measure unsteady flow in the inlet duct of a centrifugal fan operating in the state of rotating stall. They suggested that two different flow regions existed; a reverse flow region near the duct wall and a forward flow region, which was not axisymmetric and rotated at a constant speed. They also found that asymmetric flow induced by rotating stall appeared to be axisymmetric when measured by a Pitot probe, and a certain correspondence existed between the time-averaged and unsteady rotating flows. Tsurusaki et al. (1987) studied experimentally the rotational speed of stall cell in vaneless diffusers and critical inlet flow angle for rotating stall without scroll. They suggested two empirical equations for the rotational speed and a prediction method for critical inlet flow angle for the onset of rotating stall.

Gui et al. (1989) performed an experimental and numerical investigation of splitter blades in a forward-curved centrifugal fan. The numerical simulations were carried out using a FEASM (Finite Element Approximate Solution Method) (Gu 1984). In this work, effects of three geometric parameters, that is length, circumferential position, and stagger angle of the splitter blades, on fan performance were studied. The results indicated that, when the splitter blades located near the suction surface of the blades, the efficiency increased significantly, but, when located near the pressure surface, the total pressure coefficient increased. The length and stagger angle of the splitter blade did not affect the fan performance. Kim et al. (2012a) also investigated the splitter blades in a centrifugal fan using 3D RANS analysis with SST turbulence model. A total of three impeller configurations (Figure 4.73) with different numbers of main blades were tested, and all the configurations had two splitter blades in each blade passage. The two splitter blades had different chord lengths; 90% and 30% chord of the main blade. They found that the centrifugal impeller with seven main blades showed the highest pressure rise and efficiency as shown in Figure 4.74.

Velarde-Suarez et al. (2001) carried out an experimental investigation of a forward-curved blade centrifugal fan. At two radial locations of the impeller exit, steady velocity components and unsteadiness levels were measured using a hot-wire anemometer. At the two impeller exit radial locations considered, a large asymmetry in circumferential velocity distribution was found with considerable changes in both magnitude and direction of the velocity vector, especially at the volute tongue as shown in Figure 4.75. Regarding the velocity unsteadiness distribution and velocity components power spectra, in the vicinity of the volute tongue, the high levels were observed at a low flow rate (Figure 4.76). Following this research, Ballesteros et al. (2002) performed an experimental study to investigate turbulence intensity in the

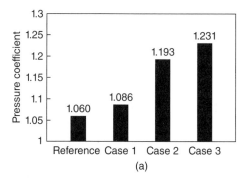
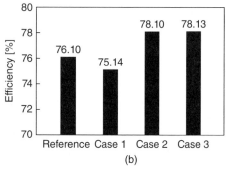

Figure 4.74 Comparison of the performance parameters of a centrifugal fan. (a) Pressure coefficient and (b) efficiency. Source: Reprinted from Kim et al. (2010a), with permission from Elsevier.

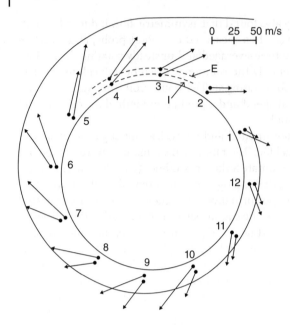

Figure 4.75 Measurement points and absolute velocities at impeller outlet of a forward-curved blades centrifugal fan. Source: Reproduced with permission from Velarde-Suarez et al. (2001). Copyright © 2001 by American Society of Mechanical Engineers.

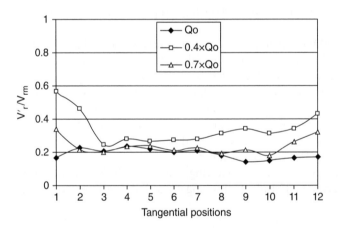

Figure 4.76 Circumferential distribution of velocity unsteadiness in the radial direction at radial location $R_1 = 1.05R_2$ in a forward-curved blades centrifugal fan. Source: Reproduced with permission from Velarde-Suarez et al. (2001). Copyright © 2001 by American Society of Mechanical Engineers.

centrifugal fan by using the same experimental facility. The results showed that the turbulence intensity was high in the vicinity of the volute tongue at lower flow rates.

Thakur et al. (2002) developed two quasi-steady rotor-stator models, called the frozen interface model and averaged interface model, to analyze the flow field in a centrifugal blower and to predict performance using 3D RANS analysis. In the frozen interface model, the information between the rotor and stator was exchanged locally, and the averaged interface model averaged circumferentially the variables on the rotor-stator

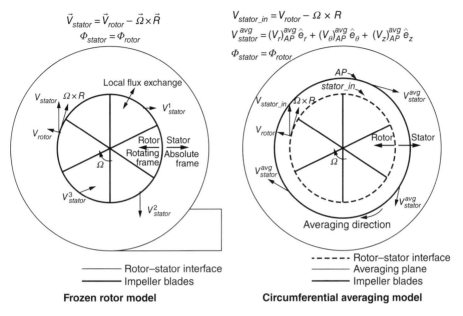

$$\vec{V}_{stator} = \vec{V}_{rotor} - \vec{\Omega} \times \vec{R}$$
$$\Phi_{stator} = \Phi_{rotor}$$

$$V_{stator\_in} = V_{rotor} - \Omega \times R$$
$$V_{stator}^{avg} = (V_r)_{AP}^{avg}\,\hat{e}_r + (V_\theta)_{AP}^{avg}\,\hat{e}_\theta + (V_z)_{AP}^{avg}\,\hat{e}_z$$
$$\Phi_{stator} = \Phi_{rotor}$$

Frozen rotor model   **Circumferential averaging model**

Figure 4.77 Illustration of two rotor-stator models in a centrifugal blower. Source: Reproduced with permission from Thakur et al. (2002), ASCE.

interface as presented in Figure 4.77. The static pressure rise, horsepower, and static efficiency were predicted by both models. For the static pressure rise, both models underpredicted the value at lower mass flow conditions and overpredicted the value at higher mass flow conditions compared to experimental data. The frozen interface model predicted the horsepower well, whereas the averaged interface model underpredicted the horsepower. The static efficiency was overpredicted by the averaged interface model in an entire range of mass flow rate. On the other hand, the frozen interface model underpredicted the efficiency at lower mass flow condition but overpredicted it at higher mass flow conditions, as presented in Figure 4.78.

Chen et al. (1996) proposed a numerical model to simulate the 3D flow in a forward-curved (or multiblade) centrifugal fan using RANS analysis. Their numerical model included modeling of so-called "blade force," which replaces the role of rotating blades in the impeller and thus simulates the changes in velocity and pressure through the impellor. Thus, with this model, the blade-to-blade flow in the impeller is not calculated, and without specifying the blades steady flow through the impellor could be simulated by putting the "body force" model in the source terms of the momentum equations in the impeller block. The merit of this approach was to reduce the computational meshes (memory) in the impeller block and thus the computational time compared to the Navier–Stokes analysis of the whole domain. Because only the circumferential force changes the momentum and the total energy of the flow in the impeller, modeling of the circumferential force, which uses an empirical formula for a velocity coefficient suggested by Eck (1975), was included in the "blade force" model. Later, Seo et al. (2003) modified this model and validated the numerical model using experimental data obtained by Kim and Kang (1997) for a forward-curved centrifugal fan. In this model, the radial force in the impeller was also considered and

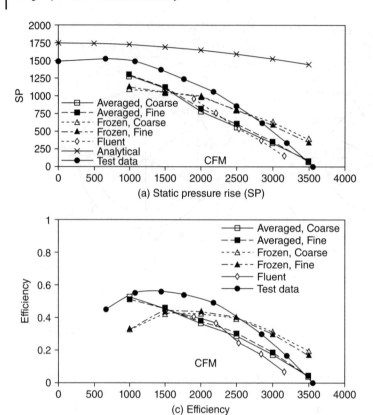

Figure 4.78 Comparison with experiment of performance data obtained with different models for a centrifugal blower. Source: Reproduced with permission from Thakur et al. (2002), ASCE.

the experimental data without scroll were used to find optimum velocity coefficients and efficiencies. The "body force" models for circumferential (f_c) and radial (f_r) forces are as follows;

$$f_c = \frac{\dot{m}[d_2(d_2\omega/2 - c_{2r}\cot\beta_2)\varepsilon - d_1 c_{1u}]}{\bar{d}} \tag{4.100}$$

$$f_r = \frac{1}{2}\bar{A}\{c_{2u}[(1+\eta_{im})u_2 - c_{2u}] - c_{1u}[(1+\eta_{im})u_1 - c_{1u}]\} - \sum \frac{\Delta V \rho}{r}c_u^2 \tag{4.101}$$

where $\dot{m}$ is mass flow rate, d is diameter of impeller, $\bar{d}$ is average of the inlet and outlet impeller diameters, ω is angular velocity of impeller rotation, β is outlet flow angle, ε is slip factor, u is blade speed, $\bar{A}$ is average of the inlet and outlet areas of the impeller, r is radius, ρ is fluid density, η_{im} is efficiency of impeller, ΔV is cell volume, and c_r, and c_u are radial and tangential velocity components, respectively. And, subscripts, 1 and 2 indicate the inlet and exit of the impeller, respectively. Figure 4.79 shows the grid system in the impeller block, a diagram of forces acting on an individual cell, and velocity triangles. The numerical results for velocity, pressure, and flow angle at the impeller exit agreed reasonably well with the experimental data, and the 3D flow structure in the scroll was reproduced well by the model.

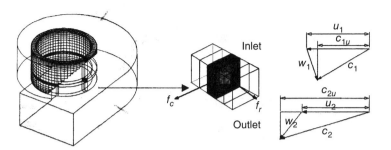

Figure 4.79 Grid system of the impeller block, diagram of forces acting on the cell, and velocity triangles for a forward-curved blades centrifugal fan. Source: Reproduced with permission from Seo et al. (2003), Copyright © 2003 by Institution of Mechanical Engineers.

Guo and Kim (2004) proposed a slip factor model improved from the conventional model (Yamazaki 1986, 1987a,b) and a correction method for numerical calculation of the flow through the impeller in a forward-curved centrifugal fan. To validate the developed slip factor model and correction method, they carried out steady and unsteady 3D RANS analyses with a k-ε turbulence model. The improved slip factor (μ) model for the forward-curved blade impeller was given by

$$\mu = \frac{C'_{u2}}{C_{u2}} = \frac{C_{u2} - \Delta C_{u2}}{C_{u2}} = 1 - \frac{\frac{U_2}{D_2}a + \frac{Q}{4zb_2R_b}}{U_2 - \frac{Q}{\pi D_2 b_2 tan\beta_2}} = 1 - \frac{a/D_2 + \varphi\pi D_2/4zR_b}{1 - \varphi/tan\beta_2} \quad (4.102)$$

where C_u is absolute circumferential velocity component, ω is rotating speed, a, and b are effective passage width and blade width, respectively, U is peripheral speed, D is diameter, Q is volume flow rate, z is blade number, R_b is radius of curvature of blade, φ is flow coefficient, and β is blade angle. Subscripts 1 and 2 indicate impeller inlet and outlet, respectively. The prime value indicates actual flow velocity obtained from slip factor model. The mass-averaged absolute circumferential velocity was expressed as follows;

$$\overline{C}_{u2} = C'_{u2} + \Delta C'_{u2} \quad (4.103)$$

where the following expression for the correction was suggested in this work.

$$\Delta C'_{u2} = w'_2 \cos(\pi - \beta'_2)\frac{\varepsilon}{1 - \varepsilon} \quad (4.104)$$

where

$$\varepsilon = B_F + (1 - B_F)B_s$$
$$\beta'_2 = \pi - \tan^{-1}\frac{C_{m2}}{w'_{u2}} \quad (4.105)$$

Here, ε is total blockage coefficient, C_m is meridional velocity, w is relative velocity, and B_F and B_S are blockage coefficients due to backflow near the front plate and due to flow separation from the suction surface, respectively. Prime values indicate actual flow angle and velocity obtained from the slip factor model. The improved slip factor model with the correction method predicted accurately the mass-averaged absolute circumferential velocity at the exit of the impeller near and above the flow rate of peak total pressure coefficient as shown in Figure 4.80.

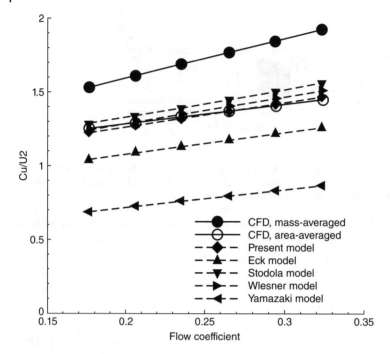

Figure 4.80 Comparison of predicted circumferential-averaged velocity at exit of impeller in a forward-curved blades centrifugal fan. Source: Reproduced with permission from Guo and Kim (2004), Copyright © 2003 by American Society of Mechanical Engineers.

Khelladi et al. (2005) investigated the flow at the impeller-diffuser interface of a vaned centrifugal fan through experimental and numerical studies. They focused on the effect of an axial gap between the impeller and the casing on the flow. Numerical analysis was performed using 3D RANS equations with SST turbulence model. The results showed that the modeling that considered the axial gap in the numerical simulation showed a better prediction of local and overall behaviors of the flow in the centrifugal fan at the operating point. Figure 4.81 shows pressure-flow rate curves at the impeller inlet and the return channel outlet. The numerical results agree well with the measurements at the outlet of the return channel. At the impeller inlet, the numerical results without consideration of the axial gap deviate largely from the experimental and numerical results with an axial gap in a flow rate range from 21 to $40\,\mathrm{l\,s^{-1}}$.

Yu et al. (2005) performed a numerical study on effects of the blade inlet angle and the gap between the impeller and the inlet on the performance of a backward-curved blade centrifugal fan using 3D RANS analysis with standard k-ε turbulence model. The results indicated that the blade inlet angle and the impeller gap greatly affected the fan performance. Tajadura et al. (2006) carried out a numerical calculation to evaluate the unsteady flow in a backward-curved blades centrifugal fan using 2D and 3D RANS analyses with the k-ε turbulence model. The unsteady flow was solved using a sliding mesh technique (Gonzalez et al. 2002). Figure 4.82 shows the comparison among 2D and 3D RANS analysis results and experimental data for total pressure coefficients. Not only were the performance curves predicted well compared to experimental data, but so also was the behavior of the pressure fluctuations at various measurement points. Karanth

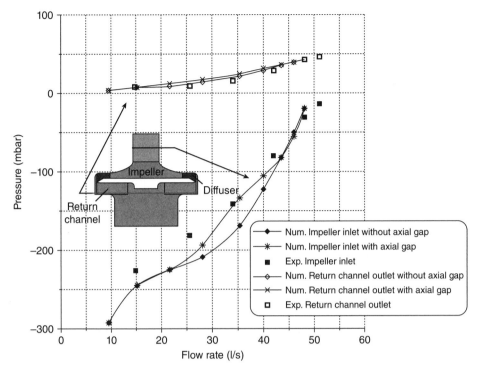

Figure 4.81 Aerodynamic characteristics of a vaned centrifugal fan. Source: Reproduced with permission from Khelladi et al. (2005), Copyright © 2005 by American Society of Mechanical Engineers.

and Sharma (2009) performed a numerical investigation of a backward-curved blade centrifugal fan with a diffuser. The analysis was focused on the effects of the radial gap between impeller and diffuser on the flow interaction and also on the aerodynamic performance of the fan using 2D URANS analysis and k-ε turbulence model. The centrifugal fan consisted of an inlet region, impeller, vaned diffuser, and volute casing. Six configurations with different radial gaps were tested in this work. The results indicated that a radial gap ratio of 0.15 showed the highest efficiency and head coefficient among the tested radial gaps.

Liu et al. (2008) conducted a numerical investigation of a centrifugal fan by solving 3D RANS equations with a Spalart–Allmaras turbulence model. The centrifugal fan consists of an inlet duct, an impeller that has 12 blades, a diffuser, and a volute. It was found that a smooth linkage between the inlet duct and the impeller inlet decreased the loss due to flow separation and enhanced the aerodynamic performance of the fan. In the case with a straight shroud, an improved performance of the fan could be obtained by selecting a reasonable range of inclined angle of the shroud, but otherwise the performance decreased as shown in Figure 4.83.

A parametric study was performed by Singh et al. (2011) for a centrifugal fan with the forward and backward-curved blades designed for cooling an automotive engine using 3D RANS analysis with a realizable k-ε model. The experiment was also carried out to measure the flow and power consumed by the fan. The performance parameters were the efficiency, power coefficient, and flow coefficient. The geometric parameters

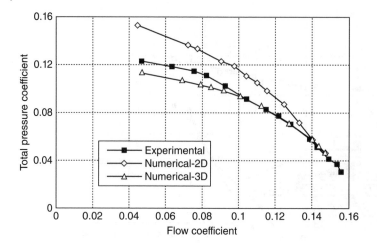

Figure 4.82 Comparison between numerical and experimental performance curves of a backward-curved blades centrifugal fan. Source: Reproduced with permission from Ballesteros-Tajadura et al. (2006), Copyright © 2006 by American Society of Mechanical Engineers.

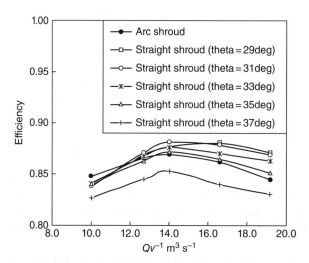

Figure 4.83 Effects of the straight shroud with different inclined angles on the efficiency of a centrifugal fan. Source: Reproduced with permission from Liu et al. (2008), Taylor & Francis LLC.

tested in this study were a number of blades, outlet angle, and diameter ratio (i.e., ratio of inner to outer blade tip length from the center). The results indicated that all fans with a different number of blades showed similar performances under HP coefficient, and, as the number of blades increased, the flow coefficient and efficiency increased (Figure 4.84). Forward-curved blades fan showed 4.5% lower efficiency with 21% higher mass flow rate and 42% higher power consumption in comparison with a backward-curved fan. Also, the experiment showed that engine temperature drop was significant with forward-curved blades leading to an insignificant effect on fuel consumption. Thus, they recommended the use of the forward-curved fan when

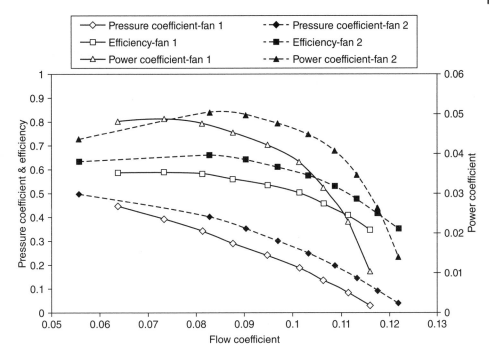

Figure 4.84 Performance characteristics comparison between centrifugal fans with 12 blades and 22 blades at 4000 rpm. Source: Reproduced with permission from Singh et al. (2011), IJAET.

vehicles require higher cooling performance. The best fan efficiency was found at a diameter ratio of 0.5.

Chunxi et al. (2011) investigated the influence of extended blades in an impeller with unchanged volute on the performance of a centrifugal fan with airfoil blades through experimental and numerical analyses. Performances of two larger impellers with increments in impeller outlet diameter of 5 and 10%, were compared with that of the original impeller without extension. The numerical calculations were carried out by using 3D RANS equations with a k-ε turbulence model in the whole flow domain. The numerical results showed that more volute loss occurred in the fan with the larger impeller. The results obtained by the experiment for extended impeller indicated that the flow rate, total pressure rise, shaft power, and SPL increased with a decrease in the efficiency in comparison with the original impeller (Figure 4.85).

Kim et al. (2013) introduced an annular plate in impeller of a forward-curved blade centrifugal fan. A parametric study of the impeller was performed to find the effect of the annular plate (Figure 4.86) on efficiency using 3D RANS analysis with an SST turbulence model. The height of the annular plate (h), the exit angle of blades, and the angle between the upper and lower impellers (θ) were chosen as the geometrical parameters tested in the parametric study. It was found that efficiency was highest when the annular plate installed at 25% span from the hub (Figure 4.87) and the numbers of blades in the upper and lower impellers are the same. The angle between the upper and lower impellers and the exit angle of blades did not significantly affect the efficiency.

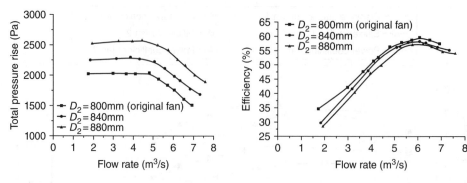

Figure 4.85 Effect of impeller enlargement on fan performance curves of a centrifugal fan with airfoil blades. Source: Reprinted from Chunxi et al. (2011), with permission from Elsevier.

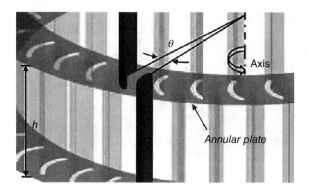

Figure 4.86 Geometric parameters related to the annular plate in the impeller of a centrifugal fan. Source: With kind permission from Springer Science+Business Media, © 2013, 1589–1595, Kim et al. (2013).

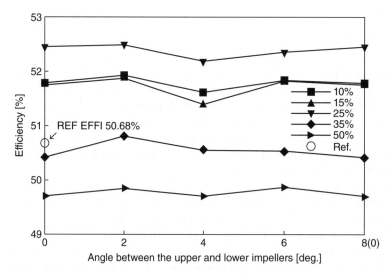

Figure 4.87 Efficiency variations with the annular plate height and the angle between the upper and lower impellers in a forward-curved blades centrifugal fan. Source: With kind permission from Springer Science+Business Media, © 2013, 1589–1595, Kim et al. (2013).

The flow in a centrifugal fan with a ported diffuser designed for vacuum cleaners was investigated experimentally and numerically by Li (2009). Numerical analysis was performed using 3D URANS analysis with the *k-ε* model. Detailed flow structures and quantitative energy losses in every step along the flow path were reported, and design improvements and applications of motor-fan system architecture were suggested.

The aeroacoustic analysis was also performed for centrifugal fans. Jeon and Lee (1997) evaluated a prediction method to identify the noise source from a centrifugal fan. Unsteady flow field and unsteady force fluctuations were analyzed by a discrete vortex method to predict the sound field. It was found that the broadband noise of an annular type centrifugal fan was caused by the unsteady force fluctuation around the impeller blades. Perie and Buell (2000) performed a study to compare the implicit numerical solution with the explicit solution for aerodynamic/aeroacoustic analysis of a forward-blade centrifugal fan. The implicit solution was a classical CFD solution with an implicit solver, a fixed topology with a rotating reference frame, and a turbulence model. The explicit solution involved an explicit solver, which accounts for topology changes and has an advantage in unsteady flow simulation as well as aeroacoustic analysis. The global behaviors of their solutions were not significantly different and only 12% difference between the implicit and explicit solutions was found for the reaction moment and the pressure jump through the fan.

Younsi et al. (2007) performed a numerical and experimental investigation into the effect of geometrical parameters on the aerodynamic and aeroacoustic performances of a forward-curved centrifugal fan. Four different impeller configurations in the same volute casing shown in Figure 4.88, were considered. These configurations show different blade spaces (VA160D with irregular spacing), blade numbers, and impeller outlet diameters. The governing equations for the numerical simulations were unsteady RANS equations with an SST turbulence model. Also, Ffowcs Williams–Hawkings equations (Williams and Hawkings 1969) were used to predict the acoustic pressure at far field. The results showed that the configuration with a smaller impeller outlet diameter (VA150) presented the best efficiency and the configuration with fewer blades (VA160E) gives the highest SPL.

4.3.4 Optimization Problems and Algorithms Used for Fan Optimization

With the recent development of high-speed computers, design optimization techniques have become more practical for turbomachinery design replacing expensive experimental approaches. Also, development of surrogate modeling promotes the application of optimization techniques to aerodynamic turbomachinery designs coupled with RANS analysis, as reviewed by Samad and Kim (2009). Figure 4.89 shows a surrogate-based optimization procedure for single- and multi-objective problems

4.3.4.1 Axial-Flow Fans

Although some optimization methods were applied to fan design using approximate 1D aerodynamic analysis (Zhou et al. 1996), a pioneering work in combining CFD (more specifically 3D RANS analysis) and optimization techniques for design of an axial-flow fan was performed by Choi et al. (1997). Most of the optimization work for fluid dynamic design reported before this work used inviscid flow or 2D viscous flow analysis instead of high-fidelity flow analysis. They applied numerical optimization techniques to the

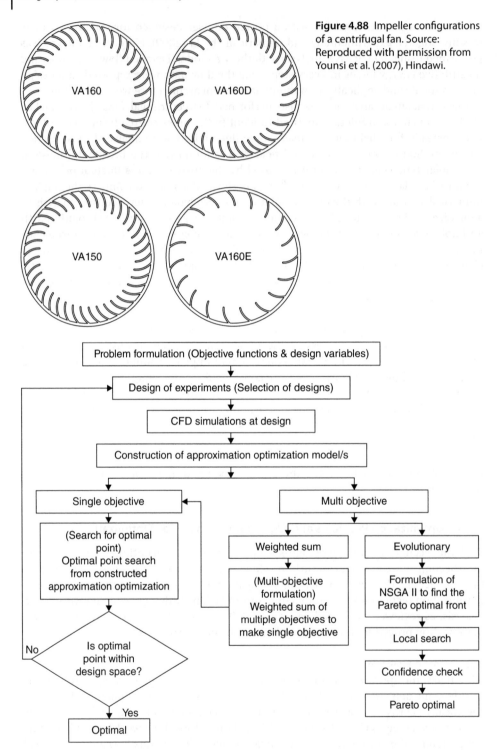

Figure 4.88 Impeller configurations of a centrifugal fan. Source: Reproduced with permission from Younsi et al. (2007), Hindawi.

Figure 4.89 Surrogate-based optimization procedure. Source: Reproduced with permission from Samad and Kim (2009), IJFMS.

Figure 4.90 Automotive cooling fan model.
Source: Reproduced with permission from Choi
et al. (1997), SAE.

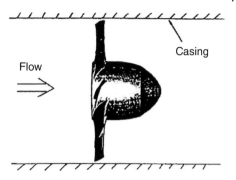

Flow

Casing

design of an automotive cooling fan (i.e., a propeller fan is shown in Figure 4.90) using 3D
RANS analysis. The conjugate gradient method (Arora 2004), which is a gradient-based
optimization algorithm, was used to search for the optimum design. Also, in order to
minimize the objective function along the search direction, the golden section method
was also used. Blade sweep was optimized to enhance fan performance, and the amount
of sweep angle change was defined by a quadratic distribution function as follows;

$$\gamma = aR_n^2 + bR_n + c \tag{4.106}$$

$$R_n = 0, \quad \gamma = 0$$
$$R_n = 0.5, \quad \gamma = \gamma_m$$
$$R_n = 1, \quad \gamma = \gamma_l \tag{4.107}$$

where,

$$R_n = \frac{(R - R_{HUB})}{(R_{TIP} - R_{HUB})} \tag{4.108}$$

Here, R_{TIP}, and R_{HUB} are radii of blade tip and hub, respectively. From this distribution
function, two design variables, γ_m and γ_t, shown in Figure 4.91 were defined; γ_m is a
sweep angle at midspan, and γ_t is that at the tip. In this work, two different objective
functions were tested separately with single-objective optimization; one is the increase
in pressure coefficient through the fan (Case 1), and the other is a ratio of the production rate of turbulent kinetic energy to the pressure head (Case 2). The latter was a quite
interesting objective function, which had not been used as a performance parameter

Figure 4.91 Definition of design variables for a propeller fan blade.
Source: Reproduced with permission from Choi et al. (1997), SAE.

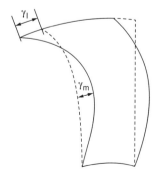

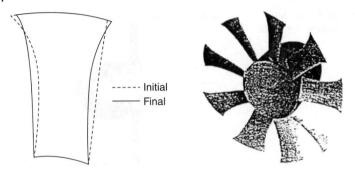

Figure 4.92 Optimized fan shape in Case 1. Source: Reproduced with permission from Choi et al. (1997), SAE.

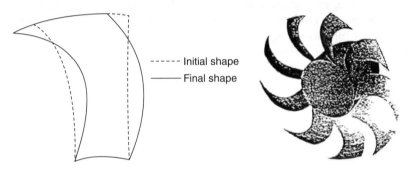

Figure 4.93 Optimized fan shape in Case 2. Source: Reproduced with permission from Choi et al. (1997), SAE.

of fans. Optimization using the former objective function (Case 1) resulted in about 17% increase in the pressure rise with the blade shape shown in Figure 4.92. On the other hand, the optimum design obtained in Case 2 (Figure 4.93), which shows large sweep of the blades, was very successful in reducing the noise level. It was proved in performance tests that the noise was reduced by about 4.5 dB in a range of flow rates, 1550–2150 m^3 h^{-1} compared to the reference fan (i.e., initial design) in Case 2, while there was no improvement in the noise in Case 1. The design shown in Figure 4.93 was employed as an automotive cooling fan by Kia Motors. However, the relationship between the production rate of turbulent kinetic energy and fan noise has not been discovered theoretically so far, and thus further study on this topic is needed.

Sorensen et al. (2000) optimized an axial fan based on the approximated aerodynamic analysis method (an arbitrary vortex flow model) suggested by Sorensen et al. (2000), which was introduced in Section 4.3.3.1. This optimization was performed in a design interval of flow rate rather than at a design point using the sequential quadratic algorithm. The objective function was defined as the fan efficiency averaged over the design interval. The hub radius of the rotor and the spanwise distributions of chord length, stagger angle from the rotational axis, and camber angle of the airfoils were selected as the design variables. They suggested that the optimum efficiency was not very dependent on the design interval for narrow design intervals, and thus their optimization method is appropriate for the design of the fans that operate in a limited range of flow rate. Sorensen (2001) also tried an optimization to minimize the trailing edge noise of the axial fan using the same aerodynamic analysis model (Sorensen et al. 2000) and the trailing edge noise model proposed by Fukano et al. (1977). He found that a large noise reduction could be obtained with a small reduction in the efficiency through the optimization.

Lotfi et al. (2006) performed an optimization of an industrial fan blade to enhance the fan efficiency using 3D RANS analysis and a GA for global optimization. They used CFD solver "MULTIP" of Denton [12] and the grid generator "STAGEN" for steady-state RANS calculation of the flow through the fan. The optimization was carried out by changing the blade camber line, lean, and sweep with constant mass flow rate and blade thickness. The camber line was modified using an eighth order Bézier curve, and the blade lean and sweep were considered using a polynomial function to adjust the increment to each blade section as a percentage of the hub chord. Through the optimization, the efficiency was improved by more than 1%. The results indicated that the efficiency was not affected significantly by the lean in the direction of rotation and backward sweep for the baseline blade, but the forward sweep improved the efficiency slightly by 0.6%.

Yang et al. (2007a) performed an optimization of an axial fan with skewed blades using an optimization algorithm based on GA and back-propagation artificial neural network (ANN) and 3D RANS analysis. The objective function was defined as a linear combination of efficiency and total pressure rise. Figure 4.94 shows their optimization procedure. By optimizing the stacking line of a radial blade, a skewed blade with the forward-skewed angle of 6.1° was obtained as the optimum blade. Through the optimization, the fan efficiency, total pressure rise, and stable operating range were increased by 1.27, 3.56, and 30%, respectively, and the aerodynamic noise was also reduced by more than 6 dB(A) at the design condition. These results were proved experimentally by Yang et al. (2008).

Chen et al. (2011) performed an optimization of a bionic fan. The fan blade was designed using the bionic method; the point cloud of the wing of the long-eared owl was scanned using a 3D coordinate measurement machine, and the fan blade shape was built by surface fitting to wing point cloud using a spline function method. The Taguchi method (Ross 1996) was used for the optimization to minimize SPL and maximize mass flow rate. The Taguchi method is a statistical quality control technique where the levels of controllable factors are selected to nullify the variation in responses due to uncontrollable factors, and thus different from the traditional design optimization techniques using a search algorithm. Among the geometric parameters, that is blade number, boss ratio, and blade stagger angle, the stagger angle was the most effective parameter. The optimum combination of the fan parameters was determined using verification experiments. The optimum fan geometry showed that the mass flow rate and the SPL were increased by 31.4% and decreased by 12.8%, respectively, compared to the initial fan.

4.3.4.2 Axial-Flow Fans

Lee et al. (2008) optimized a low-speed axial fan to enhance the efficiency of the fan using 3D RANS analysis with the SST turbulence model and RSA surrogate (Myers and Montgomery 1995) model-based optimization. The stacking line, as well as blade profile, was modified to maximize the objective function; that is, blade total efficiency. The geometric parameters, blade lean, maximum thickness, and location of maximum thickness, were selected as the design variables for the optimization. The straight stacking line was inclined by blade lean (angle), and the lean was defined as the movement of

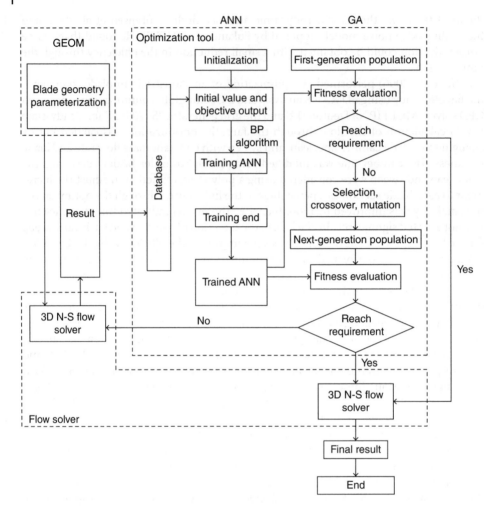

Figure 4.94 Optimization design system based on GA and back-propagation ANN. Source: Reproduced with permission from Yang et al. (2007a), Hindawi.

airfoil normal to the chord line. Through the optimization, the efficiency was increased by 1.5%. The results indicate that the total efficiency was strongly dependent on the lean of blade stacking line. In the optimized fan, the streamlines near the suction surface of the fan blade show the movement of separation lines toward the downstream direction compared to the reference fan as shown in Figure 4.95, which reduces the losses and hence increases the efficiency. For the same fan, Seo et al. (2008) optimized the blade stacking line. To maximize the fan efficiency, four geometric parameters defining span-wise distributions of sweep and lean of the blade stacking line were selected as design variables. They used the definition of lean and sweep of the blade shown in Figure 4.96. To approximate the objective function, the RSA surrogate model was used. Through the optimization, the fan efficiency was increased up to 1.75% compared with a reference fan as shown in Figure 4.97.

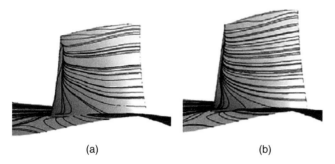

(a)   (b)

Figure 4.95 Streamlines near the suction surface of a low-speed axial fan blade: (a) reference shape and (b) optimum shape. Source: With kind permission from Springer Science+Business Media, © 2008, 1864–1869, Lee et al. (2008).

Figure 4.96 Definition of sweep and lean of axial-fan blade. Source: Reproduced with permission from Seo et al. (2008). Copyright © 2008 by Institution of Mechanical Engineers.

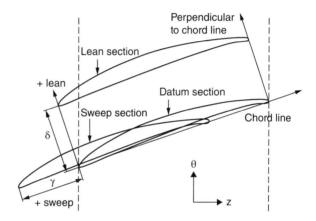

Samad et al. (2008a,b) performed multi-objective optimizations of the previously mentioned fan using a multi-objective evolutionary algorithm (MOEA) (Collette and Siarry 2003) and 3D RANS analysis. Three objectives, total efficiency, total pressure, and blade torque of the fan were considered, and two optimizations each with two objective functions were performed using four design variables same as those used by Seo et al. (2008). LHS (McKay et al. 1979) was used for the DoE to select design points in the design space to construct the RSA models. For the multi-objective optimization, non-dominated sorting of GA (NSGA-II) (Deb et al. 2002) enhanced with a local search strategy was used to find Pareto-optimal fronts (Figure 4.98). Kim et al. (2010) performed another multi-objective optimization of the same axial fan for different design variables. The total efficiency and the torque were selected as two conflicting objective functions, and six parameters related to the blade lean angle and the blade profile were selected as design variables. The NSGA-II (Deb et al. 2002) with local search was used to find POS shown in Figure 4.99.

Kim et al. (2011) performed a multi-objective optimization of a ventilation axial fan using the same aerodynamic analysis and optimization methods as those in a previous study (Kim et al. 2010). The geometry, computational domain, and grid system for numerical analysis are shown in Figure 4.100. In this work, the optimization processes

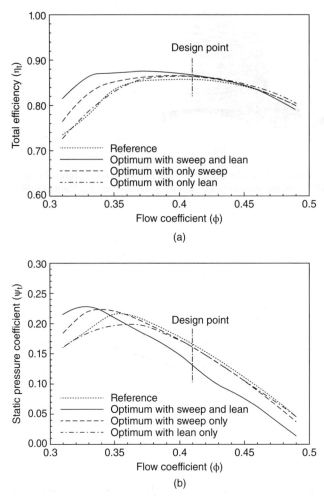

Figure 4.97 Comparison of performance and efficiency curves between optimum and reference blade shapes in an axial fan: (a) total efficiency and (b) static pressure. Source: Reproduced with permission from Seo et al. (2008). Copyright © 2008 by Institution of Mechanical Engineers.

were repeated twice to reflect the effects of diverse design variables on the two objective functions, that is the total efficiency and total pressure rise of the fan with reduced computing time. The first optimization employed three design variables defining stagger angles at the hub, midspan, and tip, and the second optimization was performed with five design variables; that is, the hub-to-tip ratio, hubcap installation distance, hubcap ratio, and the stagger angles at the midspan and tip. Definitions of these design variables are shown in Figure 4.101. The results of both the multi-objective optimizations including the POS, the RANS evaluations at some representative POS, and a reference design are shown in Figure 4.102. Because the objective functions were to be maximized, the Pareto-optimal fronts show a convex shape. Every Pareto-optimal solution has its own optimized conditions for the objective functions. Extreme ends of a Pareto-optimal front represent a pair of extreme optimum solutions; the highest value of one objective

Figure 4.98 Pareto optimal solutions for efficiency and total pressure of an axial fan. Source: Reproduced with permission from Samad et al. (2008a), World Scientific Publishing.

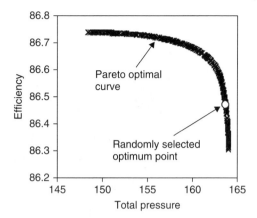

Figure 4.99 POS by hybrid MOEA for an axial fan. Source: With kind permission from Springer Science+Business Media, © 2010, 2059–2066, Kim et al. (2010).

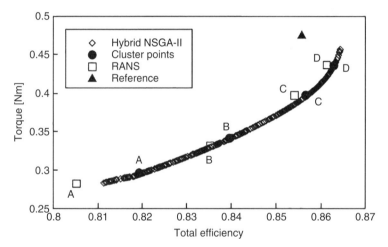

function and the lowest value of the other. The POS values of the second optimization (E–I) are shown to be better for both objective functions than those of the first optimization (A–D). The relative errors of the objective functions predicted by the RSA model were less than 1.0% compared to the RANS analysis results. The highest increments in the efficiency obtained from the optimizations were 0.0115 and 0.0182, respectively, for the first and second optimizations compared with that of the reference model without a hubcap at 0.7421. The highest pressure rises for the first and second optimizations were 110.264 and 112.239 Pa, respectively. These are much larger than the value of the reference design at 85.433 Pa. However, these performance improvements are partly attributed to the installation of the hubcap.

Kim et al. (2014) performed another multi-objective optimization to simultaneously enhance the aerodynamic and aeroacoustic performance of this ventilation axial fan. 3D steady and unsteady RANS equations with the SST turbulence model were used for aerodynamic analysis and, based on the results of unsteady flow analysis, the Ffowcs Williams–Hawkings equations were solved for aeroacoustic analysis. Single- and multi-objective optimizations were carried out sequentially. The total efficiency was

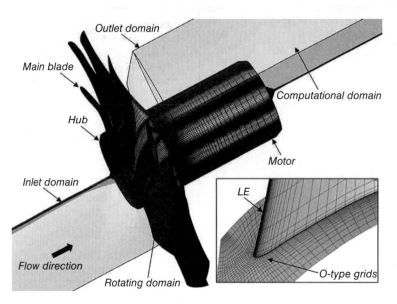

Figure 4.100 Three-dimensional geometry and computational domain of an axial-flow fan. Source: Reproduced with permission from Kim et al. (2011). Copyright © 2011 by American Society of Mechanical Engineers.

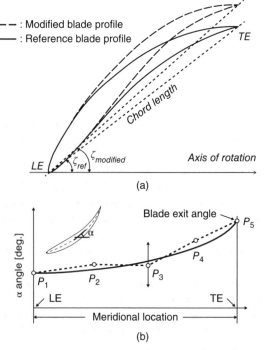

Figure 4.101 Definition of design variables for an axial fan: (a) Definition of the stagger angle, (b) angle distribution according to a Bézier curve, (c) blade generation by interpolation with B-spline curve, (d) schematic diagram defining the hub-to-tip ratio, and (e) meridional view defining the hubcap. Source: Reproduced with permission from Kim et al. (2011). Copyright © 2011 by American Society of Mechanical Engineers.

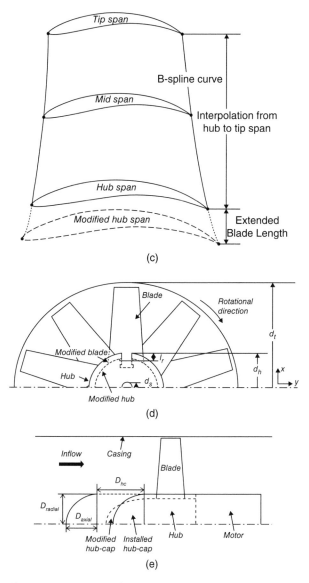

(c)

(d)

(e)

Figure 4.101 (*Continued*)

employed as an objective function for single-objective optimization. The objective function was approximated by a PRESS (predicted error sum of squares)-based-averaging (PBA) model (Goel et al. 2007) with five design variables; the hub-to-tip ratio, the hub-cap installation distance, the hubcap ratio, and the angle distributions at the midspan and tip of a blade. The multi-objective optimization was performed based on the result of the single-objective optimization to simultaneously increase total efficiency and reduce the overall SPL using a hybrid MOEA coupled with an RSA surrogate model. Two design variables defining the sweep and lean angles at the blade tip, which were not tested in the single-objective optimization, were used in multi-objective optimization.

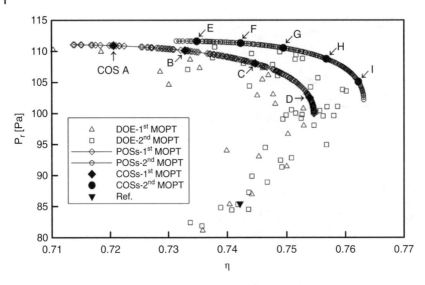

Figure 4.102 Results of the first and second multi-objective optimizations (MOPTs) for an axial fan. Source: Reproduced with permission from Kim et al. (2011). Copyright © 2011 by American Society of Mechanical Engineers.

Through the single-objective optimization, the total efficiency was increased by 1.90% compared to a reference design, but its overall SPL was also increased by 0.47 dBA. Among the POSs obtained through the multi-objective optimization, a noise-oriented design, AOD 1, and an efficiency-oriented design, AOD 2, showed 1.44% and 2.20% enhancements in the total efficiency, respectively, and 0.62 and 0.44 dBA reductions in the overall SPL, respectively, when compared with the reference design as shown in Figure 4.103. Therefore, the multi-objective optimization further improved both the aerodynamic and aeroacoustic performances by optimizing the sweep and lean angles of the blades.

A tunnel ventilation jet fan was optimized by Kim et al. (2012b) based on 3D RANS analysis and surrogate modeling. Figure 4.104 shows the computational domain and grid system for the analysis of the jet fan, which consists of a casing, silencer, electric motor and rotor, and stator blades. PBA model (Samad et al. 2008a,b) was used as the surrogate model to approximate the objective function; that is, the total efficiency. Four geometric parameters defining the meridional length and thickness profiles of the rotor blade at the hub and shroud were employed as the design variables for optimization. The objective function values were calculated by 3D RANS analysis at 35 design points generated by LHS for the four design variables. Also, the PBA model was constructed based on these objective function values as follows;

$$Fwt, avg = 0.584 FRSA + 0.396 FKRG + 0.020 FRBNN \qquad (4.109)$$

where F_{RSA}, F_{KRG}, and F_{RBNN} represent the RSA, KRG, and RBNN models, respectively. The PBA model is a weighted-average of these basic surrogate models and is expected to protect against choosing a poor surrogate model. Through the optimization the total efficiency was enhanced by 1.11% compared to the base model.

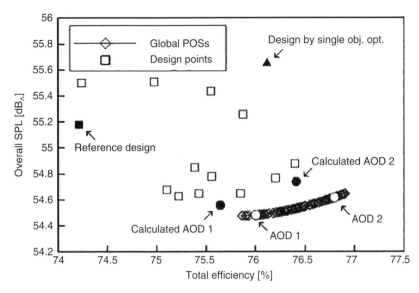

Figure 4.103 Global POS for an axial fan. Source: Reproduced with permission from Kim et al. (2014). Copyright © 2014 by the American Institute of Aeronautics and Astronautics, Inc.

Aulich et al. (2013) performed a multidisciplinary optimization of a counter-rotating fan. This work considered different objectives in aerodynamic, mechanic, aeroelastic, and manufacturing aspects. In the counter-rotating integrated shrouded propfan, both rotors were already aerodynamically optimized in a first design phase. A new optimization strategy was developed to find a rig-ready design. In the optimization, two objective

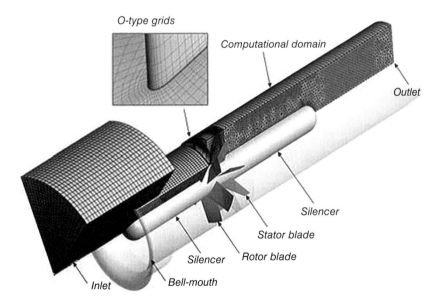

Figure 4.104 Computational domain and grid system of a jet fan. Source: With kind permission from Springer Science+Business Media, © 2012, 1793–1800, Kim et al. (2012a).

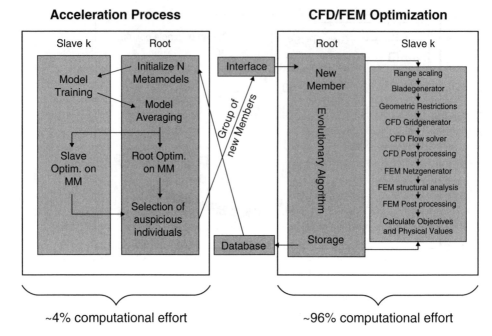

Figure 4.105 Flowchart of optimization. Source: Reproduced with permission from Aulich et al. (2013). Copyright © 2013 by American Society of Mechanical Engineers.

functions were considered with 106 design variables and a large number of aerodynamic and mechanical constraints, which are related to four profiles of the rotor blade at different blade heights and the axial positions of the blades. The mechanical behavior of the fan blades was improved through four successive aeromechanical optimizations. Until the third optimization, to secure the improvement obtained in the previous optimization, the objective functions become constraints in the next optimization. Since the aerodynamic performance decreased slightly in this process, the fourth optimization was carried out additionally to enhance the efficiency with maintaining the achieved mechanical behavior. The optimization procedure is shown in Figure 4.105.

4.3.4.3 Centrifugal Fans

Han et al. (2003) and Han and Maeng (2013) performed similar optimizations of a forward-curved blade centrifugal fan using RSA and a neural network algorithm, respectively, based on 2D RANS analysis to enhance volume flow rate (i.e., objective function) in terms of the angle and a curvature radius of cutoff. To reduce the computing time, they used 2D RANS analysis to evaluate aerodynamic performance. To obtain boundary conditions at the inlet of the impeller for the 2D calculations, they performed an experiment for the fan with various angles and curvatures of the cutoff. The inlet boundary conditions were obtained at the 60% impeller width where the flow is not affected by the inactive zones based on the experimental observation that the inactive flow zones due to flow recirculation are formed at angles of cutoff between 150° and 180° and within about 50% impeller width from the entrance of the fan. Figure 4.106 shows that there are considerable differences in the volume flow rate between the

Figure 4.106 Comparison of volume flow rate in a forward-curved blades centrifugal fan. Source: Reprinted from Han et al. (2003), © 2003, with permission from Taylor & Francis Ltd.

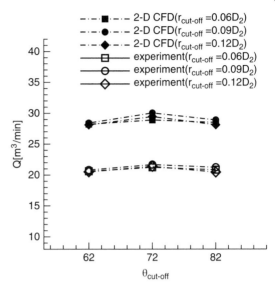

experimental measurement and the 2D RANS analysis, but the qualitative behaviors are quite similar. Han et al. (2003) obtained the optimum design at a cutoff angle of 72.4° and a cutoff curvature radius of 0.092 times the outer diameter of the impeller, where a separation around the cutoff was diminished.

Kim and Seo (2004) optimized a forward-curved blade centrifugal fan using 3D RANS analysis and an RSA surrogate model. To reduce the large computing time due to a large number of blades in the forward-curved blade fan, they used the impeller force models proposed by Seo et al. (2003) where the action of the moving blade (blade force) is modeled by a body force introduced to each computational cell in the impeller block, and regarded the flow inside the fan as steady flow. The objective function of the optimization was fan efficiency, and the location of cutoff (θ_c), the radius of cutoff (R_c), expansion angle of the scroll (α), and width of the impeller (b) shown in Figure 4.107 were used

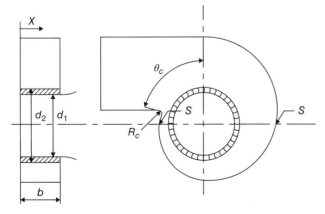

Figure 4.107 Geometry of a forward-curved blades centrifugal fan. Source: Reproduced with permission from Kim and Seo (2004), Copyright © 2004 by American Society of Mechanical Engineers.

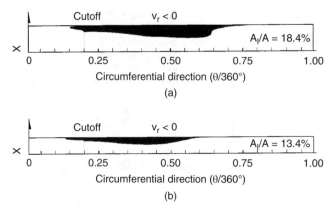

Figure 4.108 Inactive zones at exit of the impeller of a forward-curved blades centrifugal fan for (a) reference and (b) first optimum shapes. Source: Reproduced with permission from Kim and Seo (2004), Copyright © 2004 by American Society of Mechanical Engineers.

as design variables. Forty-two design points were selected in the design space using a 3D-optimal design in order to calculate the objective function values, which were used to construct the RSA model. The optimizations were performed with and without constraint for the fan static pressure, and the efficiency was improved effectively compared to a reference fan. As shown in Figure 4.108, a reduction of impeller width reduced the inactive zone (flow recirculation zone) from 18.4 to 13.4% of the impeller exit surface area. From analyses of computational results, they suggested that the stronger 3D flow in the scroll was found at the higher level of total efficiency, but at the lower level of static efficiency. Kim and Seo (2006) performed a similar optimization of the same fan by employing an operating parameter, that is, the flow coefficient, as a design variable instead of the expansion angle of the scroll. Because the maximum efficiency point is generally changed by optimizations, employing the flow coefficient as one of the design variables was expected to be effective in enhancing the efficiency. By the optimization, the fan efficiency was improved by 3.1% compared to the optimization with fixed flow rate as expected as shown in Figure 4.109.

Sugimura et al. (2008, 2009) proposed a design approach so-called MORDE (multi-objective robust design exploration), where multi-objective optimization techniques and data mining techniques are combined for the design of a backward-curved centrifugal fan used in a washer-dryer. Their optimization method introduced a probabilistic representation of design variables, and Kriging models were used to approximate multiple design objectives as shown in the optimization procedure (Figure 4.110). A GA optimized the mean and standard deviations of the responses (objective functions); that is, the fan efficiency and turbulent noise level. 3D steady RANS analysis was used to evaluate the objective function values. They showed that a design candidate can be selected from the non-dominated solutions through the process of analyzing and balancing the mean and variance of the performances. They also tried to obtain design knowledge by applying data mining techniques. The self-organizing map was utilized for visualization and reuse of the high-dimensional design data. Decision tree analysis and rough set theory were employed to extract design rules to enhance the performance.

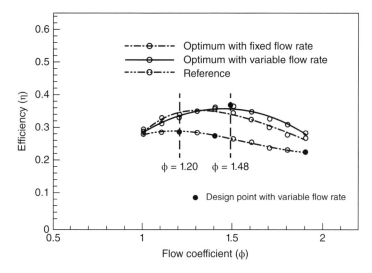

Figure 4.109 Comparison of efficiency curves between reference and optimum forward-curved blades centrifugal fans. Source: Reproduced with permission from Kim and Seo (2006), JSME.

Multi-objective optimization of a centrifugal blower with a vaned diffuser was performed by Sugimura et al. (2010). This work improved the aerodynamic efficiency and stability of the centrifugal fan by using combined methods of multi-objective optimization and quantitative design rule mining. The centrifugal impeller shape was defined by a meridional impeller profile and several blade profiles (blade sections) using a non-uniform rational B-spline (NURBS) curve. Figure 4.111 shows the meridional profile of the impeller including the vaned diffuser and the connecting vaneless diffuser. For aerodynamic analysis, 3D RANS equations with the standard k-ε model were solved. At the interface between rotating and stationary regions, a mixing plane was used as an interfacial technique. At the mixing plane, flow properties were averaged only in the circumferential direction and thus spanwise non-uniform inflow to the diffuser could be considered. Fan efficiency and root mean square of the incidence angle distribution at the vaned diffuser inlet were employed as the objective functions, and 16 parameters related to the impeller and blade profiles were selected as design variables with a constraint for the shaft power for optimization. First, the impeller was optimized using a multi-objective GA (MOGA) to enhance the fan efficiency and flow uniformity. A compromise solution was experimentally shown to improve both the objective functions. Second, decision tree analysis (Witten and Frank 2005) and rough set theory (Pawlak 1982) shown in Figures 4.112 and 4.113, respectively, were used to obtain design rules for improving each objective function. Figure 4.114 represents the design rules obtained from the decision tree analysis and correlation analysis. The rule toward the extreme design for the efficiency indicates a move from P1 to P2 in the objective function space. And, the rule toward the extreme design for the aerodynamic stability indicates a move from P1 to P3. Quantitative rules are also available for these moves. The rule set for the trade-off control that is possible by applying the qualitative rules is an example of moving from P3 to P2. The results of data mining revealed that the inlet blade angle was most important for the efficiency and the impeller exit height

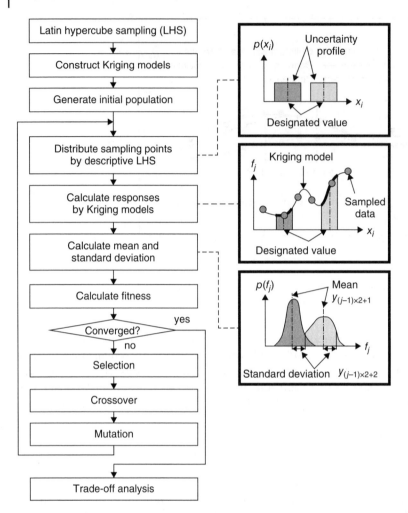

Figure 4.110 Flowchart of MORDE (*p*: probability density function, x_i: i-th design variable, f_j: j-th evaluation function, y_*:*-th objective function). Source: Reproduced with permission from Sugimura et al. (2009), JSME.

and the outer radius on the shroud side were most important for the aerodynamic stability.

Lee et al. (2011) optimized a centrifugal fan with a double-inlet impeller and double-discharge volute using 3D RANS analysis. This work used numerical optimization and experiential steering techniques to redesign the impeller blades, inlet duct, and shroud of the impeller. The objective of the optimization was to reduce the input power with a specified output pressure at the lift-side volute exit. 2D blade profile optimization was performed using a GA optimization scheme, and the impeller efficiency was improved from 92.6% (baseline design) to 93.7%. Also, blade trailing-edge shape control effectively reduced the power (from 0.945 to 0.896 reference power) while maintaining efficiency. Measurements of lift pressure coefficient were compared

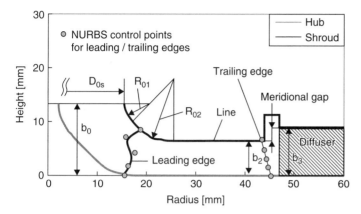

Figure 4.111 Meridional profile definition for a centrifugal blower. Source: Reprinted from Sugimura et al. (2010), © 2010, with permission from Taylor & Francis Ltd.

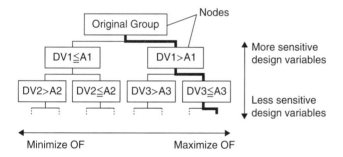

Figure 4.112 Decision tree diagram. Source: Reprinted from Sugimura et al. (2010), © 2010, with permission from Taylor & Francis Ltd.

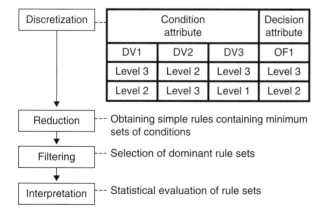

Figure 4.113 Application procedure of rough set theory. Source: Reprinted from Sugimura et al. (2010), © 2010, with permission from Taylor & Francis Ltd.

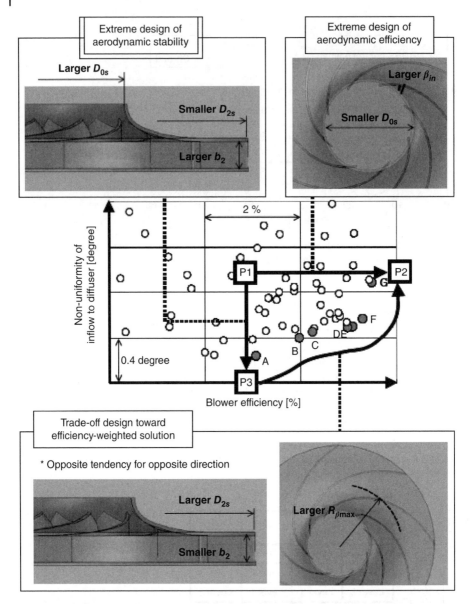

Figure 4.114 Summary of design rules based on decision tree analysis and correlation analysis. Source: Reprinted from Sugimura et al. (2010), © 2010, with permission from Taylor & Francis Ltd.

with the requirement and CFD predictions for the two existing impellers (B#1 and B#2) and a new impeller as shown in Figure 4.115.

Khalkhali et al. (2011) performed a multi-objective optimization of forward-curved blade centrifugal fans. The aerodynamic analysis was performed based on 3D RANS equations and a standard k-ε model. The objectives of this optimization were to increase the head rise and to reduce the head loss in a set of the centrifugal fan. The design variables were the number of blades and three parameters required in defining blade camber

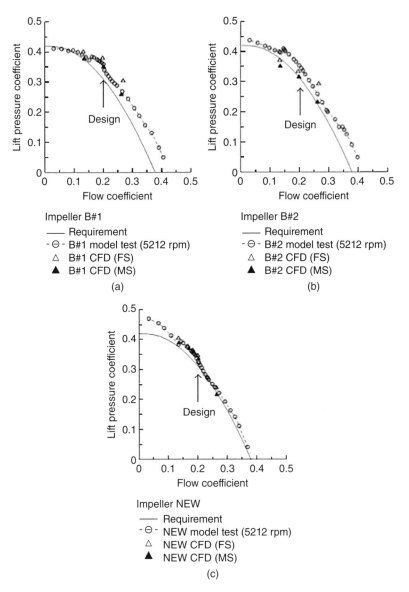

Figure 4.115 Measured lift pressure coefficient of a centrifugal fan compared with the requirement and CFD predictions for the (a) B#1, (b) B#2, and (c) new impellers. Source: Reproduced with permission from Lee et al. (2011), Hindawi.

line; leading edge angle, trailing edge angle, and the stagger angle. Two surrogate models (meta-models) were constructed as second-order polynomials of the design variables from neural networks of evolved group method of data handling (GMDH) (Farlow 1984) type. The number of numerical analyses needed to construct the surrogate models was 132 for four design variables. Finally, MOGAs were used to find POS for the two conflicting objectives, head rise and head loss.

An optimization was performed to reduce the vibration and noise of a forward-curved blade centrifugal fan based on 3D aeroacoustic analysis by Lu et al. (2012). Based on the results of 3D unsteady RANS and LES analysis, a parametric finite element analysis (FEA) model of the volute was constructed using the pressure fluctuations at blade passing frequency on the volute as external excitation forces. The FEA model was validated using experimental analysis. The thickness of the front, side, and back panel of the volute were used as design variables. To reduce the computational time for the aeroacoustic analysis during each optimization iteration, the quadratic sum of the nodal velocities instead of vibroacoustic simulation results was used as an objective function, and vibroacoustic simulation was used to validate the optimization results. Using the direct boundary element method, it was found that both the vibration and the radiated power of the vibroacoustics of the volute were reduced significantly through optimization.

Heo et al. (2015a) performed a multi-objective optimization of a centrifugal fan with splitter blades to simultaneously enhance efficiency and pressure rise. Three-dimensional RANS analysis with an SST turbulence model was used for aerodynamic performance evaluation and the hybrid MOEA and RSA surrogate models were used for the optimization. The location of the splitter and the height ratio between inlet and outlet of impeller were selected as design variables. Among the Pareto-optimal designs obtained from the optimization, two arbitrary designs, AOD1 and AOD2, showed the efficiency, and pressure rise improved by 3.81% and 69.59 Pa, and 3.82% and 63.7 Pa, respectively, compared to the design without splitters.

4.4 Hydraulic Turbines

4.4.1 Introduction

Hydraulic turbines convert the hydraulic energy stored in reservoirs or flowing rivers into mechanical energy. As inward flow reaction turbines, Francis turbines have developed since J. B. Francis (1909) invented the original hydraulic turbine in the USA in the middle of the nineteenth century. The Kaplan turbine was first developed as an axial-flow turbine in the early twentieth century. And, Pelton developed an impulse turbine in the late nineteenth century. In Francis and Kaplan turbines, a pressure drop occurs in both the rotor and stator, but no pressure drop occurs in the runner in Pelton turbines. The degree of reaction of hydraulic turbines is defined as the ratio of the static pressure drop across the runner to that across the stage. The degrees of reaction of the Kaplan and Francis turbines and pump turbines are about 90, 75, and 50%, respectively.

The normalized velocity (K_c) and the normalized circumferential velocity (K_u), which are the typical performance parameters of hydraulic turbines, are defined as follows:

$$K_c = \frac{C}{\sqrt{2gH}} \quad \text{and} \quad K_u = \frac{U}{\sqrt{2gH}} \tag{4.110}$$

where C is the flow velocity, U is the circumferential blade speed, g is gravitational acceleration, and H is the turbine head.

As described by Drtina and Sallaberger (1999), hydraulic turbines are classified by several criteria, such as shaft orientation (horizontal or vertical axis), specific speed (high,

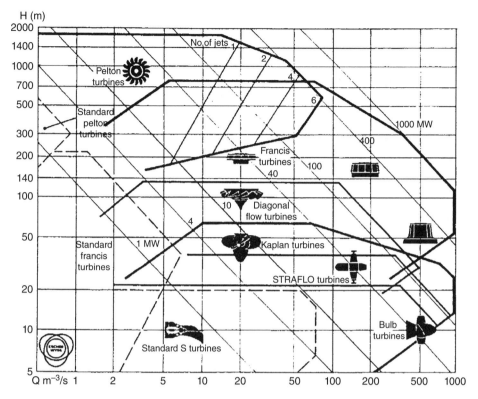

Figure 4.116 Overview of turbine runners and their operating regimes. Source: Reproduced with permission from Drtina and Sallaberger (1999), Copyright © 1999 by Institution of Mechanical Engineers.

medium or low), operation head (high pressure 200 m < H < 2000 m, medium pressure 20 m < H < 200 m, and low pressure H < 20 m), type of regulation (single, variable stator vanes, e.g., Francis; double, variable runner, and stator vanes, e.g., Kaplan or variable needle stroke and variable number of jets), and design concepts (single-stage or multistage, single-volute or double volute, and single-jet, or multi-jet). Figure 4.116 shows the types of turbines depending on volumetric flow rate (Q) and head rise (H) and includes the information on the power output. The hydraulic turbines are also classified in terms of specific speed and head rise as shown in Figure 4.117. The number of runner vanes generally reduces as specific speed increases.

Analysis using CFD became popular in the design of hydraulic machinery in recent decades. Keck and Sick (2008) reviewed extensively the development of CFD methods applied to hydraulic turbomachinery. They divided the whole development period into four sub-periods: The potential flow and Q-3D period (1978–1987), the 3D-Euler period (1987–1994), the steady RANS period (1990–2000), and unsteady, multiphase, and multiphysics period (2000).

In the first period, the flows through the turbomachinery blades that deviates from potential flows were calculated using a quasi-3D approximation of the 3D Euler equations by introducing stream functions both in the meridional and the blade-to-blade analysis.

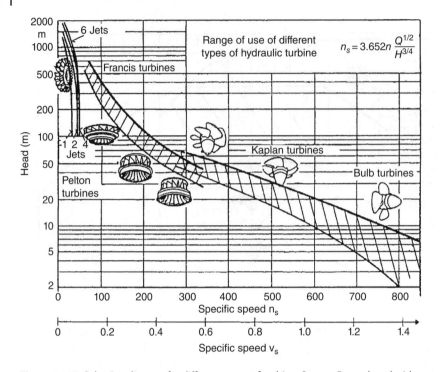

Figure 4.117 Selection diagram for different types of turbine. Source: Reproduced with permission from Drtina and Sallaberger (1999), Copyright © 1999 by Institution of Mechanical Engineers.

In the second period of CFD development, the analysis using 3D Euler equations was dominant. Göde and Rhyming (1987) reported on the first successful 3D Euler analysis in a Francis runner. The off-design calculation with high accuracy was possible through 3D Euler analysis. They presented Figure 4.118 which shows a leading-edge vortex simulated numerically for the first time by using 3D Euler analysis at off-design conditions.

In the period from 1990 to 2000, the steady RANS equations became dominant tools for the analysis of hydraulic machinery. The finite volume methods, which are conservative with respect to mass and momentum, were used to solve the RANS equations in the hydraulic turbines. In contrast to the Euler equations, the RANS equations consider both turbulent and viscous effects. The hill chart of a Francis turbine was first presented using CFD by Keck et al. (1996). This CFD-based hill chart agreed well with that obtained by experiment as shown in Figure 4.119.

From 2000 to present, the most remarkable developments were found in the unsteady RANS analysis of two-phase flow and fluid-structure interaction. Unsteady phenomena are usually found between the rotating and stationary parts of turbomachinery. To account for the effect of cavitation in the performance prediction, two-phase flow simulation is necessary. The Rayleigh–Plesset approach is used for modeling the formation and collapse of vapor bubbles. A comparison between computational and experimental vortex rope in draft tube is shown in Figure 4.120 (Stein et al. 2006).

Figure 4.118 Leading edge vortex in a Francis runner modeled by the 3D Euler code. Source: Göde et al. (1989).

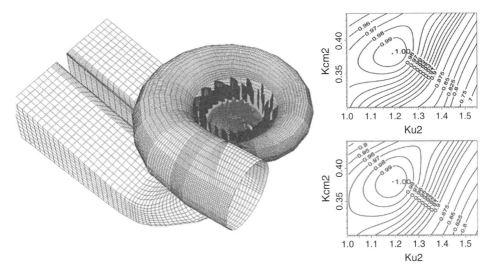

Figure 4.119 Entire Francis turbine, numerical hill chart versus experimental hill chart. Source: Keck et al. (1996).

4.4.2 Cavitation in Hydraulic Turbines

Vapor bubbles can be generated in low-pressure regions and collapsed in HP regions in hydraulic machinery. The blade surface is subjected to high local stresses through these phenomena called cavitation. The performance of hydraulic turbines can be declined after few years of operation, as the turbines get severely damaged by cavitating erosion. It is impossible to entirely eliminate the cavitation in hydraulic turbines but can be reduced to an economically acceptable level. Therefore, in this respect, the suitable measures were suggested by various researchers (Kumar and Saini 2010).

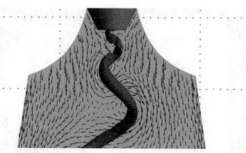

Figure 4.120 Cavitating draft tube vortex as observed at the test rig (left) and predicted by CFD (right). Source: Stein et al. (2006).

Thoma's cavitation number is an important parameter describing cavitation in hydraulic turbines, which is defined as follows (from Drtina and Sallaberger 1999):

$$\sigma = \frac{h_{at} - h_v - H_s}{H} \tag{4.111}$$

where h_{at} is the atmospheric pressure head, h_v is the vapor pressure head, and H_s is the suction pressure at the outlet of reaction turbine or height of the turbine runner above the ail water surface. In order to avoid cavitation, this cavitation number must be larger than the value of σ at which experimental tests prove the onset of cavitation.

Figure 4.121 presents the variation of turbine efficiency with Thoma's cavitation factor (σ) (Avellan 2004). When the value of Thoma's cavitation factor σ is larger than critical cavitation factor σ_o, the cavitation does not occur and the turbine efficiency is unchanged as shown in the figure. As σ decreases below σ_o with an increase in the suction head (H_s), the turbine efficiency starts to decrease after a small increase.

Various types of cavitation in Francis turbines are shown in Figure 4.122 (Kumar and Saini 2010). The generalized Rayleigh–Plesset equation was formulated to approximate the bubble growth. If the bubble pressure ($P_B(t)$) and the infinite domain pressure ($P_I(t)$) are given, it is possible to find the radius of the bubble ($R_B(t)$) from the following equation (Kumar and Saini 2010):

$$\frac{P_B(t) - P_\infty(t)}{\rho} = R_B \frac{d^2 R_B}{dt^2} + \frac{3}{2}\left(\frac{dR_B}{dt}\right)^2 + \frac{4v}{R_B}\frac{dR_B}{dt} + \frac{2\gamma}{\rho R_B} \tag{4.112}$$

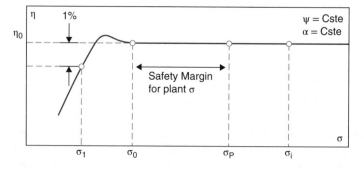

Figure 4.121 Cavitation curve of a Francis turbine for a constant machine specific energy coefficient and a given guide vane opening angle. Source: Avellan (2004).

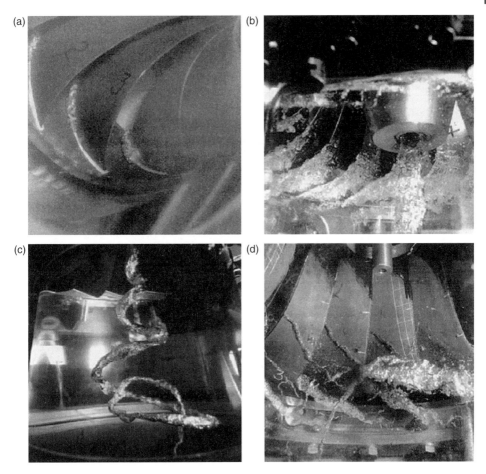

Figure 4.122 Various cavitation types in Francis turbine: (a) leading edge cavitation, (b) traveling bubble cavitation, (c) draft tube swirl, and (d) inter-blade vortex cavitation. Source: Reprinted from Kumar and Saini (2010), with permission from Elsevier.

where v is the kinematic viscosity, γ is the surface tension, and ρ is the density.

The cavitation development was explained in terms of specific speed, load, and head by Avellan (2004). In Figure 4.123, the subscripts 1 and 2 indicate the HP and low-pressure reference sections of the system regardless of the flow direction.

The mean specific energy at a cross section of the flow passage (gH) is defined as:

$$gH = \frac{P}{\rho} + gZ + \frac{C^2}{2} \tag{4.113}$$

where p is the absolute pressure, Z is the elevation of a point, and C is the mean velocity. Thus, the specific hydraulic energy E is defined as the difference in the mean specific energy between the high and the low-pressure sections.

$$E = gH_1 - gH_2 \tag{4.114}$$

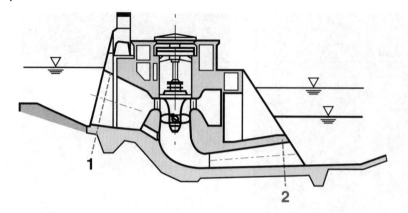

Figure 4.123 Schematic of a run-off power plant with Kaplan Turbine. Source: Avellan (2004).

The hydraulic power P_h is defined as a product of the discharge mass flow rate and specific energy

$$P_h = \rho QE \qquad (4.115)$$

Then, the overall efficiency η is defined as:

$$\eta = \frac{P_h}{P} \text{ for a pump and } \eta = \frac{P}{P_h} \text{ for a turbine.} \qquad (4.116)$$

and the dimensionless discharge and energy coefficients are defined as:

$$\varphi = \frac{Q}{\pi\omega R^3} \text{ and } \psi = \frac{2E}{\omega^3 R^2} \qquad (4.117)$$

where, Q is discharge, ω is the angular velocity, and R is the reference radius of the machine runner/impeller. The relationship between these discharge and specific energy coefficients can be plotted on a diagram called an efficiency hill chart, as shown in Figure 4.124.

According to the International Electrotechnical Commission (IEC) standard, the *NPSE* is defined as the difference between the specific energy at section 2 and the

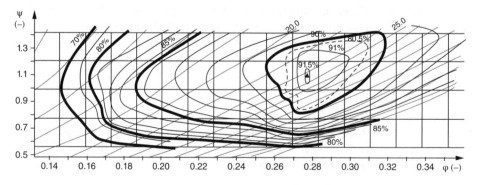

Figure 4.124 Hill chart of a Francis turbine. Source: Avellan (2004).

Figure 4.125 Typical eroded runner areas in a Francis turbine. Source: Avellan (2004).

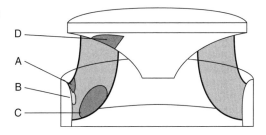

specific energy due to the vapor pressure p_v at the reference level Z_{ref}.

$$NPSE = gH_2 - \frac{p_v}{\rho} - gZ_{ref} \qquad (4.118)$$

For a turbine, assuming that all the specific kinetic energy at the turbine outlet is dissipated, *NPSE* can be approximated as follows

$$NPSE \approx \frac{p_a}{\rho} - \frac{p_v}{\rho} - gh_s + \frac{C_2^2}{2} \qquad (4.119)$$

where p_a is atmospheric pressure and h_s is machine setting, $h_s = Z_r - Z_a$. Thoma's cavitation factor σ, defined previously, can also be written in terms of *NPSE* and the specific hydraulic energy as:

$$\sigma = \frac{NPSE}{E} \qquad (4.120)$$

In the case of a Francis turbine, the cavitation development is strictly driven by the specific energy coefficient and the flow coefficient. Four typical eroded areas of a Francis runner are indicated in Figure 4.125. Severe cavitation erosions generally occur on area A.

In the case of a Kaplan turbine, a cavitation occurs at the hub of runner in the operating range. This type of cavitation is greatly dependent on the Thoma's cavitation number. Typical eroded areas of a Kaplan runner are shown in Figure 4.126. Severe cavitation erosion occurs at the blade tips and the casing.

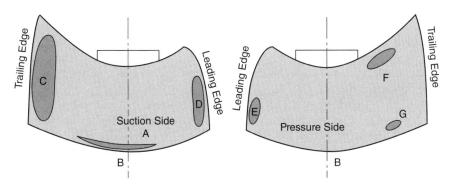

Figure 4.126 Typical eroded runner areas in a Kaplan turbine. Source: Avellan (2004).

4.4.3 Analysis of Hydraulic Turbines

Many numerical and experimental analyses have been performed for various hydraulic turbines, such as Francis and Kaplan turbines and pump turbines, which are introduced as follows.

4.4.3.1 Francis Turbines

Kurokawa and Kitahora (1994) analyzed the flow based on the theory developed by Kurokawa and Toyoura (1976), Kurokawa et al. (1978), and Kurokawa and Sakuma (1988) to determine volumetric and mechanical efficiencies in terms of specific speed for Francis turbines and Francis pump-turbines. The different types of leakage sealing and thrust-balancing devices, such as balancing pipes and balancing holes, were considered in the analysis. Sixteen types of turbine models and 12 types of pump-turbine models of different specific speeds, shown in Figure 4.127, were analyzed. The results indicated that these thrust-balancing devices had a large influence on both volumetric and mechanical efficiencies. The results for the volumetric and mechanical efficiencies are shown in Figures 4.128 and 4.129, respectively. Volumetric efficiencies of the Francis turbine and pump-turbine models were around 99%. Mechanical efficiency was around 99% for the Francis turbine model and 1% lower for the Francis pump-turbine model. In the Francis pump-turbine, mechanical efficiency in turbine mode was higher than that in pump mode.

With the development of CFD methods and computers, 3D RANS simulation of the flows in hydraulic turbines has become popular in the last couple of decades. With an introduction of the application of CFD to the analyses of various hydraulic turbines, Drtina and Sallaberger (1999) evaluated two kinds of numerical analysis methods, using 3D Euler and 3D RANS analyses, for the flow in a Francis turbine using a CFD code CFX-TASCflow. They suggested in comparison with experimental data that the relatively simple Euler method predicted the important features of the flow, but it was limited in ability to predict the losses due to the neglect of the viscous forces. The hill chart for a Francis turbine of a high specific speed was obtained using the turbine efficiency values evaluated numerically at 14 operating points with three different guide vane openings and six different heads as shown in Figure 4.130. This hill chart based on numerical calculations was in qualitatively excellent agreement with that based on experimental data. All general features were reproduced by the numerical simulations, and the BEPs were same in both the hill charts.

Susan-Resiga et al. (2003) performed a numerical study of cavitation flow in a Francis turbine. They introduced a simplified Rayleigh equation using a mixture model for two-phase cavitating flows, which was implemented in the commercial CFD code, FLUENT. This model was validated for a hemispherical fore-body cavitator. In the Francis turbine runner, the predicted cavity shape and position agreed well with the flow visualization results at the BEP.

Wang et al. (2007) performed a LES with a one-coefficient dynamic sub-grid scale (SGS) stress model for a Francis turbine. The turbulent quantities in the turbine passage were calculated. The investigation was focused on finding the structure of the distorted wakes arising from the guide vanes. The LES prediction for pressures on the pressure and suction surfaces showed good agreements with experimental data. The results indicated that the trailing wakes induced by the guide vanes had a significant influence on the turbulence structure near the blades.

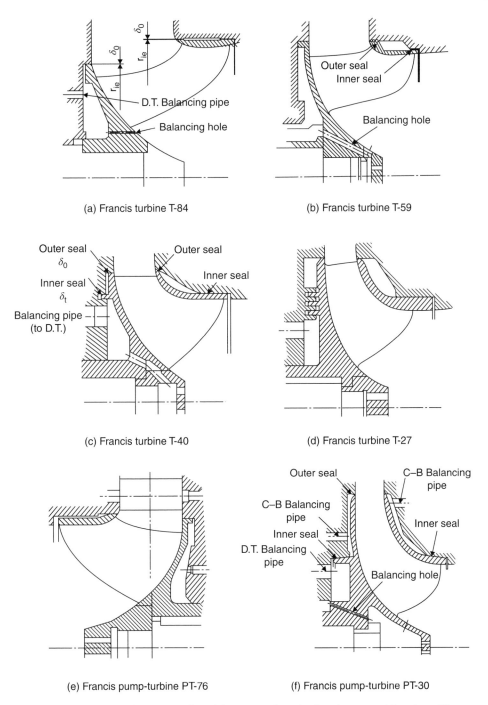

Figure 4.127 Runner configurations of model Francis turbine (a–d) and pump-turbines (e and f). Source: Kurokawa and Kitahora (1994).

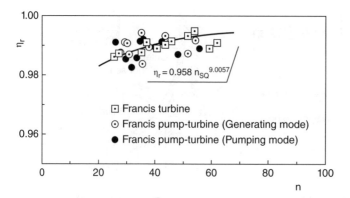

Figure 4.128 Volumetric efficiency for Francis turbines. Source: Kurokawa and Kitahora (1994).

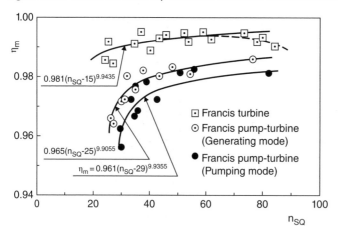

Figure 4.129 Mechanical efficiency for Francis turbines. Source: Kurokawa and Kitahora (1994).

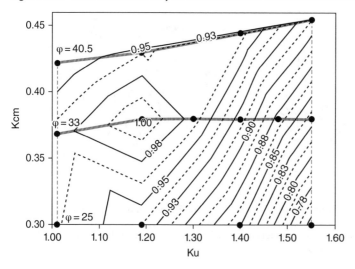

Figure 4.130 Hill chart based on 14 numerically obtained efficiency values for a Francis turbine (● data points; --- constant guide vane opening). Source: Reproduced with permission from Drtina and Sallaberger (1999), Copyright © 2008 by Institution of Mechanical Engineers.

A numerical investigation of complex flow phenomena in the draft tube of a Francis turbine was performed using 3D RANS analysis by Zhang et al. (2009). This study focused on the physical mechanisms underlying the complex flows under a part-load condition. Moreover, they presented a control methodology for eliminating undesirable helical vortex ropes. From the results, three guiding principles were proposed for effective control of the flow under the part-load condition. First, the control should be imposed at the inlet of the draft tube rather than downstream of the cone segment. Second, the control should focus on dealing with the reversed axial flow at the inlet of cone segment. Last, the control needs a sufficiently strong interference by either solid or fluid means. To test these principles, a strong axial jet within a thin cylindrical region near the axis was adopted. This jet added a mass flux to the draft tube flow but the azimuthal velocity profile remained constant. Moreover, once the transient process was over, the flow in the draft tube came in a quasi-steady state, which was ultimately desired. This means that the control using water jet was effective (Figure 4.131).

Chirkov et al. (2012) developed a hybrid 1D-3D CFD model for numerical simulation of a Francis turbine. Effects of penstock length on pressure and discharge surge were investigated using this model. The Francis turbine system was divided into two parts. The first part is the turbine itself where the unsteady 3D two-phase flow is dominant. The other part is a pipe line that can be simulated using 1D hydro-acoustic equations. The spiral case was omitted for simplicity but represented by its energy loss. This simplified pipe line system consisted of a single penstock and pipe line with the constant cross section. In the turbine domain, the cavitating flow was described as the

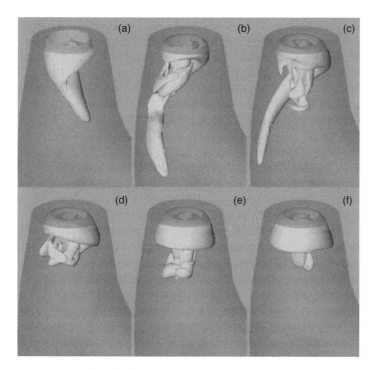

Figure 4.131 Controlled flow by jet injection in the draft tube of a Francis turbine. Starting from the onset of control, the dimensionless times in (a)–(f) are $t = 0$, 3.60, 5.04, 7.92, 12.25, and 32.4, respectively. Source: Reproduced with permission from Zhang et al. (2009), Copyright © 2009 by American Society of Mechanical Engineers.

flow of isothermal compressible liquid-vapor mixture, and the terms of evaporation and condensation were evaluated using the Singhal et al. model (1997). The results indicated that pressure pulsation decreased when the length of penstock was increased.

Wu et al. (2013) numerically investigated the cavitating flow with different openings of guide vane in a Francis turbine using 3D RANS analysis with SST k-ω turbulence model and a mixture model of cavitation. An equal critical cavitation coefficient line was calculated in the analysis. Computational results showed good agreements with experimental data. From the results, a spiral vortex rope was observed as shown in Figure 4.132 for different openings of guide vane. This spiral vortex rope was transformed into a column vortex rope as the opening increased. They suggested that, as for the equal critical cavitation coefficient line, the energy losses in the turbine were caused by the vortex rope in the draft tube and flow separation in the runner.

Pressure pulses in a Francis pump-turbine were predicted numerically under turbine operating conditions with different MGV (misaligned guide vane) openings by Xiao et al. (2014) using unsteady 3D RANS analysis with SST turbulence model. Pressure pulses in the turbine were analyzed for four MGVs with two different openings. Numerical results for the hydraulic performance and the pressure fluctuations showed that the MGVs reduced the amplitudes of relative pressure fluctuation in the stationary flow passage, not in the runner blade region. With the increase of the MGV openings, the pulse amplitude decreased and vertical motion in the blade passage was weakened.

Trivedi et al. (2013) performed numerical and experimental investigations of a Francis turbine with a high head in a whole operating range. Unsteady 3D RANS analysis was performed for five operating conditions with two turbulent closure models, that is, SST

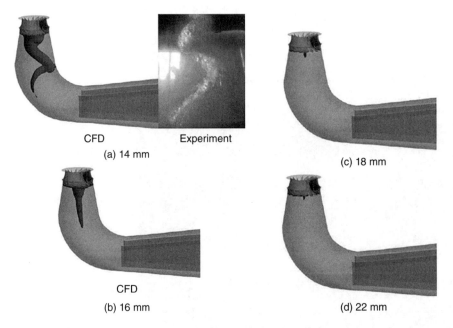

Figure 4.132 Vortex rope in draft tube of a Francis turbine for different openings of guide vane. Source: With kind permission from Springer Science+Business Media, © 2013, 1635–1641, Wu et al. (2013).

and standard k-ε models, and two schemes for convection terms, that is, high resolution and second-order upwind schemes. A Hill diagram was obtained and time-dependent pressure was measured at the design and off-design conditions by using pressure sensors installed in the rotating and stationary parts in the turbine. A comparison between numerical and experimental hydraulic efficiencies is shown in Figure 4.133, where the lowest differences in the efficiency are found at the BEP, while the maximum difference of about 11% is found at the lowest discharge. Both the numerical and experimental results for pressure-time signal showed that the rotor-stator interaction caused a torque oscillation in a particular range of power generation. Fourier analysis of the signals indicated that a vortex rope occurred in the draft tube at the off-design condition.

As well as the numerical works, experimental works were also performed actively for hydraulic turbines using advanced measuring devices. Especially, the PIV system gives the opportunity to explore experimentally the mean velocity field and the turbulence in hydraulic machinery. The flow field with cavitation vortex shown in Figure 4.134, called a turbine rope, at the outlet of a Francis turbine runner was analyzed by Iliescu et al. (2003) using PIV. Two cameras' images, one for the velocity field and the other one for the rope contour, were processed separately. A result of the analysis of the rope is shown in Figure 4.135. In addition to this study, the same authors, Iliescu et al. (2008) investigated the rope characteristics, diameter, and center position with the Thoma cavity number. Figure 4.136 shows an example of the results of image processing. It was found that the rope diameter decreased with an increase in Thoma cavity number, and the eccentricity of the rope position decreased with an increase in Thoma cavity number.

Tridon et al. (2008) performed an experimental investigation on a Francis turbine using PIV. The PIV measurements were taken in the draft tube and the conical diffuser of the turbine. This study focused on describing the swirling flow evolution in the turbine outlet as well as the phenomena that led to a sudden drop in the turbine efficiency. According to the results, the main effect of "accident" was found to increase the vorticity at the center of the flow duct. The "accident" means an undesirable efficiency drop

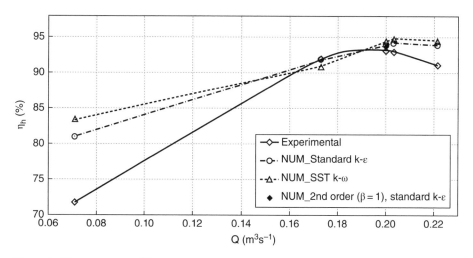

Figure 4.133 Comparison of the experimental and numerical hydraulic efficiency of a Francis turbine at five operating conditions. Source: Reproduced with permission from Trivedi et al. (2013), Copyright © 2013 by American Society of Mechanical Engineers.

Figure 4.134 Development of the rope in a Francis turbine for $\sigma = -0.370$. Source: Reproduced with permission from Iliescu et al. (2003), Copyright © 2003 by American Society of Mechanical Engineers.

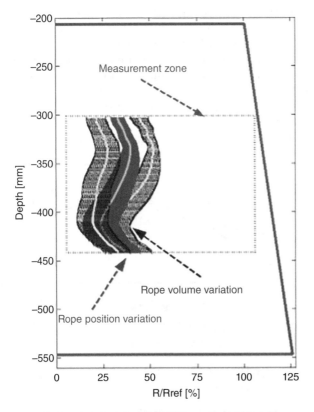

Figure 4.135 Standard deviation of the rope position and rope volume. Source: Reproduced with permission from Iliescu et al. (2003), Copyright © 2003 by American Society of Mechanical Engineers.

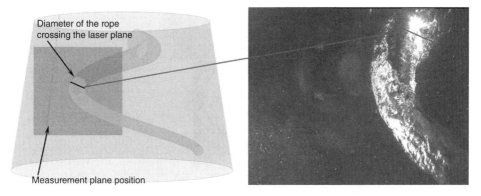

Figure 4.136 Extraction of the rope diameter by image processing from each instantaneous image. Source: Reproduced with permission from Iliescu et al. (2008), Copyright © 2008 by American Society of Mechanical Engineers.

caused by an upgraded runner as the discharge is increased above the BEP value. However, at this stage of the study, they could not establish a direct link between the swirling flow at the runner outlet and the flow rate unbalance at the draft tube outlet. The flow in the turbine runner is very complex and highly turbulent. The radial velocity is seldom measured because of a complicated measurement setup. However, measurements of velocity components are required to properly initialize the numerical study. Tridon et al. (2010) measured velocity components in downstream of the runner in a Francis turbine using LDV and PIV techniques (Figure 4.137). Also, an analytic formulation for these velocity components was compared with measurements. The velocity and the length nondimensionalized, respectively, by the runner outlet velocity and the runner outlet radius were used in this study. Three-dimensional visualizations of the measured axial and tangential velocities at four operating points are shown in Figure 4.138. They also introduced an analytical formulation of the velocity components based on the analysis of Susan-Resiga et al. (2006), where a theoretical model was suggested from a velocity profile analysis at the inlet of the draft tube. The formulations for three velocity components agreed well with the experimental data.

4.4.3.2 Kaplan Turbines

Nilsson and Davidson (2000) numerically investigated tip clearance losses in a Kaplan water turbine using 3D RANS analysis with a k-ω turbulence model. The computational domain consists of the blocks for the guide vanes and the runner, and the inlet boundary conditions for the computations in the runner block came from the circumferentially averaged values at the guide vane block. Four different operating conditions were considered in this study. The analysis results indicated that tip clearance flow reduced the turbine efficiency by about 0.5%.

Muntean et al. (2004) performed a numerical analysis to investigate the effects of the two different stay vane configurations on the stay vane loading in a Kaplan turbine using FLUENT 6.0 code. The numerical results showed that the first configuration of the vane had a very large angle of attack leading to a flow detachment. The second configuration adjusted a stay vane position with significant improvement in the local flow and the

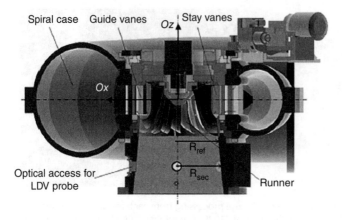

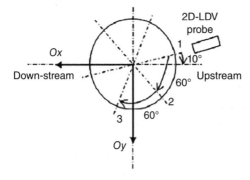

Figure 4.137 Sketch of the Francis turbine model and LDV flow survey section in the cone. Source: Reprinted from Tridon et al. (2010), with permission from Elsevier.

corresponding reduction in the vane loading. However, when the flow angle circumferential variation is considered, the second stay vane configuration actually increases the flow angle non-uniformities. Consequently, this led to the unsteady loading of the larger runner blade.

Petit et al. (2010) performed an experimental and numerical investigation on the influence of the inlet pipe curvature on the flow in a Kaplan turbine. The numerical simulations were performed using the Open FOAM CFD code. The velocity profiles were measured by using LDA techniques. The experimental results were used to validate the numerical results. The results indicated the importance of including the curved pipe in the computational domain. In comparison between the numerical results and the experimental data, the predicted velocity profiles agreed satisfactorily with measured data. However, they suggested that for the accurate flow prediction, it is essential to couple the runner with the draft tube and to compute transient simulation.

Grekula and Bark (2001) performed an experimental study of cavitation process in a Kaplan turbine to find mechanisms that promote erosive cavitation. The measurements in this work were obtained using video filming, high-speed filming, and visual observations with a stroboscopic light. This study focused on investigating the several types of

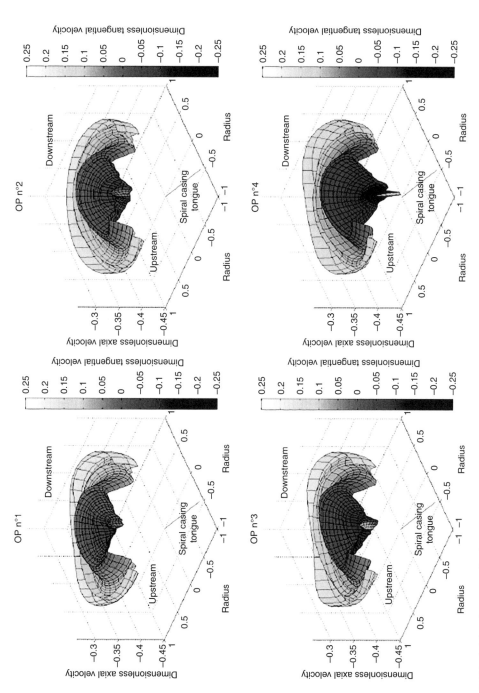

Figure 4.138 Axial and tangential velocity under the runner of a Francis turbine obtained by LDV–OP1 to OP4. Source: Reprinted from Tridon et al. (2010), with permission from Elsevier.

cavitation; those at the blade leading edge, at the blade root, and at the blade tip (all on the suction side), which were dominant and led to severe erosion on the runner.

4.4.3.3 Pump-Turbines

Pump-turbine operation in the unstable region of the characteristic is not acceptable because it may cause self-excited vibration of the system. Pump-turbines show two typical features of unstable behavior. One occurs in generating mode at low-load operation close to runaway conditions; an S-shape in the turbine characteristic. The other one occurs in pump operation as a head drop as the flow is reduced; the saddle-type instability of head curve. Figure 4.139 shows turbine characteristics fulfilling the criterion for instability. Because of the curve shape on the right, these instability characteristics are called S-shaped. During turbine startup and synchronization, these instabilities are highly undesirable. Hasmatuchi et al. (2009) performed an experimental investigation on off-design operations of a low specific speed centrifugal pump-turbine to find the onset and development of flow stability. The research focus was placed on the generating mode at off-design conditions with runaway and S-shapes. The results indicated that the pressure fluctuation was dominated by the blade passing frequency and its first

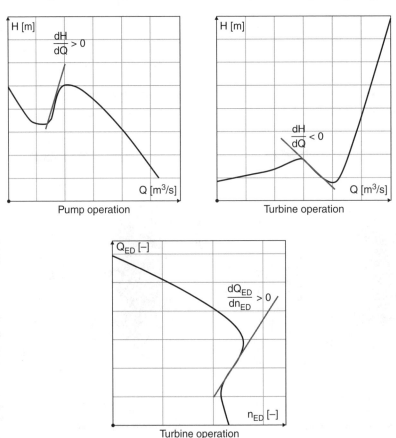

Figure 4.139 Saddle shaped pump head curve and slope of turbine characteristic at no-load for constant wicket gate angle, in dimensional, and normalized values. Source: Reproduced with permission from Genter et al. (2012). IOP Conference Series: Earth and Environmental Science, 15, no. 3, 032042, 2012, IOP Publishing.

harmonic at the BEP. Genter et al. (2012) conducted a numerical and experimental investigation of a pump-turbine to analyze the flow in runner and diffuser in the unstable operation region. They identified a mechanism for the S-shape turbine characteristic. Basic mechanisms of instability were identified using numerical analysis but with a certain offset compared to measurements in the analysis of the unstable behavior of pump at part load.

Wang et al. (2011) investigated numerical schemes in predicting S-shape performance of a pump turbine using 3D RANS with an SST turbulence model. They suggested that a second-order scheme was effective for turbine and shut-off modes and the pressure staggering option (PRESTO) model was suitable for the reversible pump mode. A comparison between numerical and experimental results for the S-shaped curve is shown in Figure 4.140, where the agreements are good except in the low-speed region. Analysis of computational flow field indicated that at three typical operating conditions different flow structures were found in the runner zone. For example, Figures 4.141 and 4.142 show the predicted vertical flow in runner channels at operating condition B indicated in Figure 4.140. The large vortex caused severe blockage to the channel and thus huge pressure fluctuations.

Yin et al. (2013) investigated an S-shaped curve in a pump turbine both numerically and experimentally. Using a hydraulic loss analysis based on previous CFD results, they found that an S-shaped curve was caused by the hydraulic loss in the runner. The S-shaped characteristic was suppressed by broadening the meridional section using the analysis of the blade loading distribution. To validate this method, two runners with different meridional sections were designed by means of the inverse design method. Through the experimental test, it was proved that the runner with a broader meridional section had a stable performance curve and the S-shaped curve was eliminated.

The MGV can improve the stability of the pump turbine in startup mode and no-load mode. The pressure fluctuation in a Francis pump-turbine was investigated numerically under turbine mode with three different MGV arrangements/openings by Xiao et al. (2012). The tested MGVs showed that the efficiency and power were increased up to 15

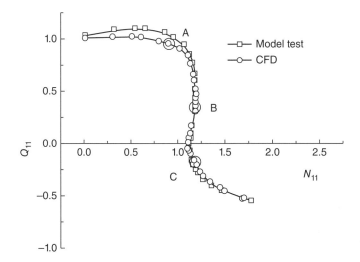

Figure 4.140 Comparison of S-shaped curve of a pump-turbine between model test and CFD results (normalized with respect to the most efficient point). Source: With kind permission from Springer Science+Business Media, © 2011, 1259–1266, Wang et al. (2011).

Figure 4.141 Relative velocity vectors on S1 in CF stream surface (B) in a pump-turbine. Source: With kind permission from Springer Science+Business Media, © 2011, 1259–1266, Wang et al. (2011).

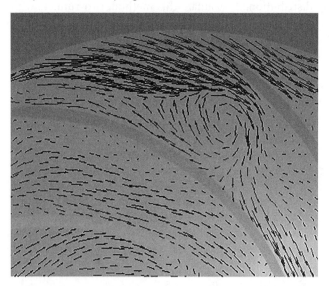

Figure 4.142 Vortex flow in runner channels of a pump-turbine. Source: With kind permission from Springer Science+Business Media, © 2011, 1259–1266, Wang et al. (2011).

and 50%, respectively, compared to the case without MGV. The pressure analysis showed that the amplitude of pressure pulse was reduced significantly by using MGV.

Numerical analysis of flow field in a pump-turbine in pump mode was performed by Braun et al. (2005) using 3D RANS analysis with an SST model. The pump-turbine model consisted of an impeller with five blades, a diffuser with 22 guide and stay vanes, and a spiral casing. The results indicated that development of the detached flow had no influence on the computed global performances. In case of the hydraulic energy distribution at the impeller outlet, the distribution became asymmetric toward lower discharge value. The hydraulic energy near the shroud was increased with decreasing discharge value whereas it decreased near the hub. The absolute minimum value of hydraulic energy was located in the middle of the channel in all cases.

Zobeiri et al. (2006) performed a numerical and experimental investigation of the rotor-stator interaction in a reduced scale pump-turbine model for the maximum discharge operating condition in generating mode. The pump-turbine model included 20 stay vanes, 20 guide vanes, and nine impeller blades. The numerical calculations were performed in the whole internal domain of the pump turbine from the spiral casing to the draft tube using 3D unsteady RANS analysis with a k-ε turbulence model. The measurements for pressure were performed using piezoresistive miniature pressure sensors in the distributor channels. The predicted pressure fluctuations showed good agreements with experimental data. The numerical results showed that the maximum pressure amplitude of blade passage frequency (BPF) occurred in the rotor-stator zone, but it reduced rapidly backward to the stay vane.

Yin et al. (2010) performed a numerical study to investigate the flow pattern in a pump-turbine at off-design conditions in pump mode. The numerical results were obtained using 3D RANS equations and the SST k-ω turbulence model. It was found from the results that "jet-wake" pattern (Figure 4.143) occurred near the band, and the special head-flow profile was caused by the special loss characteristics of the stay and guide vanes. The dimensionless hydraulic loss and total head without stay and guide vanes are shown in Figure 4.144.

4.4.4 Optimization of Hydraulic Turbines

4.4.4.1 Kaplan Turbines
A multi-objective optimization of a Kaplan turbine runner was performed by Lipej and Poloni (2000) using a CFD code, "CFX-TASCflow," and MOGA. 3D RANS equations with the k-ε turbulence model were used for the flow analysis in the Kaplan turbine. The

Figure 4.143 Schematic plot of "jet-wake" flow pattern in a pump-turbine. Source: With kind permission from Springer Science+Business Media, © 2010, 3302–3309, Yin et al. (2010).

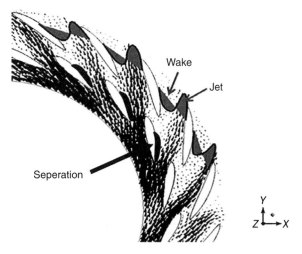

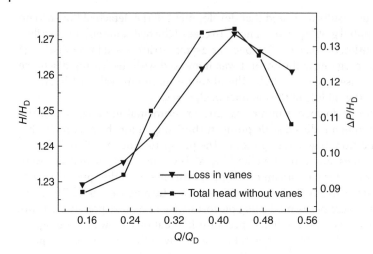

Figure 4.144 Hydraulic losses and vaneless head vs non-dimensional flow rate in a pump-turbine. Source: With kind permission from Springer Science+Business Media, © 2010, 3302–3309, Yin et al. (2010).

chord-pitch ratio, relative profile maximum thickness, and camber position of the blade were selected as design variables. The efficiency and the relative pressure number were considered to be two conflicting objective functions. The multi-objective optimization results were obtained through POS shown in Figure 4.145.

Peng et al. (2002) performed a multi-objective optimization of an axial-flow Kaplan turbine runner using a quasi-3D (Q3D) inverse computation model (Peng et al. 1998a,b) and Powell's conjugate direct method (Lootsma 1972). The blade bound circulation distribution function and the locations of the blade leading and trailing edges were selected as design variables. The blade bound circulation is a design parameter that is important in determining the runner blade shape. A circumferentially averaged mean velocity torque distribution shown in Figure 4.146 was used to specify the blade bound circulation. The cavitation coefficient (σ) and the total hydraulic loss were considered

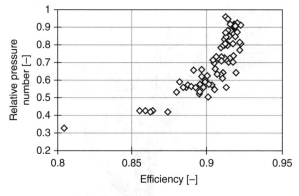

Figure 4.145 All computed configurations with Pareto front for a Kaplan turbine. Source: Reprinted from Lipej and Poloni (2000), © 2000, with permission from Taylor & Francis Ltd.

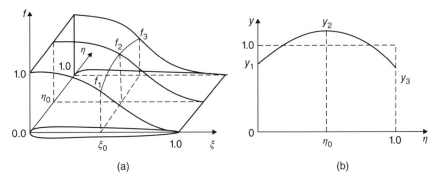

Figure 4.146 Velocity torque distribution in blade region of an axial-flow Kaplan turbine: (a) dimensionless distribution function and (b) blade duty distribution. Source: Reprinted from Peng et al. (2002) (figure 3 from original source), © 2002, with the permission of John Wiley & Sons, Inc.

conflicting objective functions. The cavitation coefficient was defined as follows:

$$\sigma = \lambda K_w^2 + \zeta_t K_v^2 \tag{4.121}$$

where

$$\lambda = \left(\frac{W_k}{W_o}\right)^2 - 1, \quad K_w = \frac{W_0}{\sqrt{2gH}}, \quad K_v = \frac{V_0}{\sqrt{2gH}} \tag{4.122}$$

Here, W_k denotes the relative velocity at the location of the lowest pressure, H the effective head, g the gravity, and W_0, and V_0 are the mean relative velocity and the mean absolute velocity at the exit of runner blades, respectively. ζ_t is the efficiency of the draft tube. The other objective function, the total hydraulic loss, was defined as follows:

$$F_\eta = 1 - \zeta(x) \tag{4.123}$$

where ζ denotes the hydraulic turbine efficiency and x denotes a vector of design variables. An optimal solution obtained from the multi-objective optimization showed 3.63 and 16.75% reductions in the cavitation coefficient and the total hydraulic loss, respectively, compared to a reference design.

Hydraulic efficiency of a Kaplan turbine was optimized by Arnone et al. (2009) using 3D RANS analysis with a CFD code, HYDROMES (Arnone 1994) for flow analysis, and ANN for the optimization algorithm. The stagger angles of the runner and the guide vanes were selected as design variables and the efficiency of the Kaplan turbine was chosen as the objective function. The optimization was performed by changing the stagger angles with various operating conditions. The predicted stagger angles agreed well with experimental data over the tested flow rate range with a maximum discrepancy of about 2°.

Banaszek and Tesch (2010) performed an optimization of a Kaplan turbine using GA with ANN and steady-state 3D RANS analysis. The polytropic loss coefficient was defined as an objective function. The blade thickness distribution of the rotor was selected as design variable and modified by using the distribution coefficient along the mean camber line. The polytropic loss coefficient of the optimized Kaplan turbine was reduced by 8.1% with constant mass flow rate.

4.4.4.2 Francis Turbines

An optimization of a Francis turbine to improve both the efficiency and the cavitation performance was carried out by Wu et al. (2007). An iterative algorithm was used to enhance the hydraulic performance of the Francis turbine. The cavitation coefficient and the peak efficiency were employed as the objective functions. The optimization was performed by changing the blade angle distribution between the leading and trailing edge until the turbine satisfies the target specifications. The optimization process used in their work is presented in Figure 4.147. The Q3D codes were used for rapid flow analysis during the optimization and a commercial CFD code, "STAR-CD," was used for 3D RANS analysis in the final step of the optimization. Figure 4.148 shows the computational meshes on runner and guide vanes for RANS analysis. Through the optimization, the peak efficiency was increased by about 2.2% and the cavitation coefficient was decreased in the whole flow range as shown Figure 4.149.

Skotak et al. (2009) presented a development process of a newly fixed blade runner called "Mixer" (Figure 4.150) of a Francis turbine throughout an optimization technique. The first step of the development was the design of the draft tube and the second step was the optimization of blade runner. They used a commercial CFD code, FLUENT 6.3, for 3D RANS analysis and the Nelder–Mead simplex method (Nelder and Mead 1965) for the optimization. The blade runner geometry was constructed with 20 independent parameters. The objective function was defined as a linear combination of the efficiency, total head, and cavitation features as follows:

$$f_o = W_E(1 - \eta) + W_H \left(\frac{|H - H_R|}{H_R} \right) + W_k \left(\sum_{P_S < P_v} (P_V - P_S) \right) \tag{4.124}$$

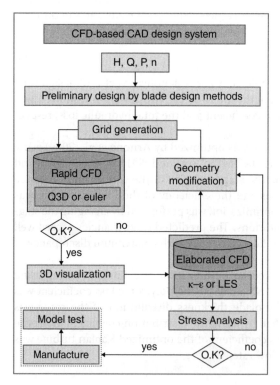

Figure 4.147 CFD-based design system. Source: Reproduced with permission from Wu et al. (2007), Copyright © 2007 by American Society of Mechanical Engineers.

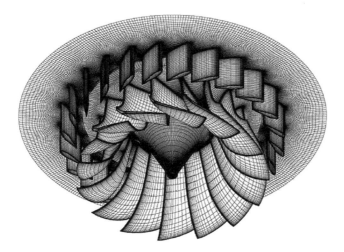

Figure 4.148 Computational meshes for a Francis turbine: surface grid of runner and guide vanes with the crown. Source: Reproduced with permission from Wu et al. (2007), Copyright © 2007 by American Society of Mechanical Engineers.

where η denotes the actual efficiency, H denotes the actual head, H_R the required head, P the actual static pressure on the blade, P_v the vapor pressure, P_s the actual static pressure on the blade, and W_E, W_H, and W_K denote the individual weighting factors. The new blade runner developed successfully through the minimization of the objective function with about 400 iterations. And, the efficiency was achieved up to approximately 90% and the head was about 5 m (Figure 4.151)

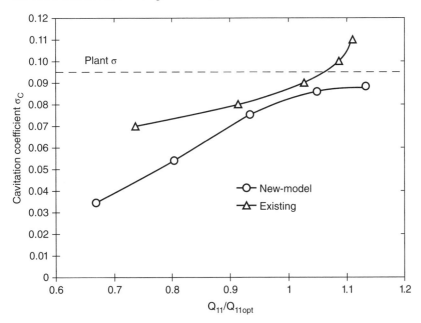

Figure 4.149 Measured model critical cavitation coefficients at a rated head of $Ht = 73$ m for a Francis turbine. Source: Reproduced with permission from Wu et al. (2007), Copyright © 2007 by American Society of Mechanical Engineers.

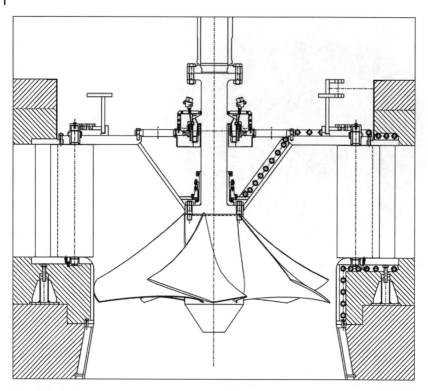

Figure 4.150 New propeller runner of a Francis turbine. Source: Reproduced with Skotak et al. (2009), IJFMS.

Because the interaction effect between a draft tube and a runner increases with the increase of the specific speed of the hydraulic turbine, the trade-off between runner loss and draft tube loss is recommended. Nakamura and Kurosawa (2009) performed a multi-objective optimization to reduce the hydraulic losses in draft tube and blade runner of a Francis turbine using 3D RANS analysis and MOGA. Thirty geometric parameters of the blade runner and 50 geometric parameters of the draft tube were selected as design variables. The 3D shape was defined using the geometrical parameters related to meridian flow passage and several cross-sectional shapes. The multi-objective optimization was conducted with 100% flow rate of the turbine. Through the multi-objective optimization, both the hydraulic losses in the draft tube and the runner were decreased. This improved the turbine efficiency in a wide operating region. Similarly, Lyutov et al. (2015) carried out a multi-objective optimization to enhance the efficiency in the wide operating range of a Francis turbine using 3D RANS analysis and MOGA. The governing equations with k-ε turbulence model were solved by using an in-house code, CADRUN (Cherny et al. 2003, 2005). They performed two optimizations; Case 1 considered only the runner, but Case 2 considered both the runner and the draft tube. Twenty-eight parameters for the runner and nine parameters for the draft tube were considered as design variables. The runner shape shown in Figure 4.152 was determined by the blade surface and meridian projections of the hub, shroud, and inlet and outlet edges of the blade. The draft tube was determined by changing its median section and the shape of

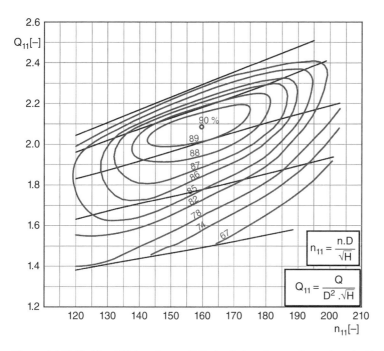

Figure 4.151 Turbine hill chart from a CFD simulation. Source: Reproduced with Skotak et al. (2009), IJFMS.

Figure 4.152 Variation of RZ-projection of a Francis runner. Source: Reproduced with permission from Lyutov et al. (2015), Copyright © 2015 by American Society of Mechanical Engineers.

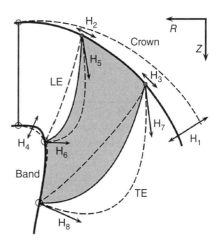

the outlet cross section as shown in Figure 4.153. The turbine efficiencies at part load and full load conditions were selected as the objective functions. The Pareto optimal solutions of the two multi-objective optimizations were compared in Figure 4.154. The average efficiency of Case 2 was 0.3% higher than that of Case 1. The results indicated that the draft tube should be considered with the runner when designing the hydraulic turbine.

Enomoto et al. (2012) performed an optimization of a Francis turbine runner using 3D RANS analysis with RNG k-ε model and a conventional optimization method with

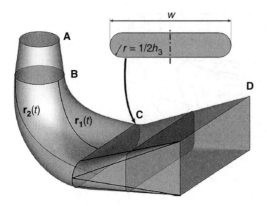

Figure 4.153 Draft tube (DT) shape construction for a Francis turbine. Source: Reproduced with permission from Lyutov et al. (2015), Copyright © 2015 by American Society of Mechanical Engineers.

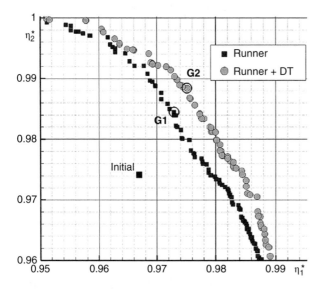

Figure 4.154 Pareto fronts for optimization runs Opt01 and Opt02 for a Francis turbine. Source: Reproduced with permission from Lyutov et al. (2015), Copyright © 2015 by American Society of Mechanical Engineers.

DOE. Computational domain for the optimization was limited to the region from runner to draft tube. Also, the velocity distribution obtained from the RANS analysis with Reynolds stress turbulence model in a flow domain from the spiral casing to draft tube was set as the boundary condition at the runner inlet. For the optimization, eight geometric parameters related to the runner shape were considered as design variables. A total weighted-average efficiency, defined as a linear combination of the turbine efficiencies at 12 operating points from partial load to over load and from high to the low head, was selected as an objective function. The pressure coefficient on the runner was used for the constraint. Through the optimization, the efficiency of the Francis turbine was improved more than the reference turbine over the whole flow rate range as shown in Figure 4.155.

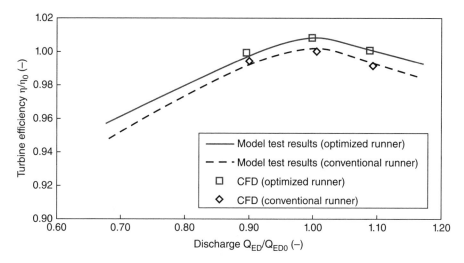

Figure 4.155 Model performance of a Francis turbine. Source: Reproduced with permission from Enomoto et al. (2012), IOP Conference Series: Earth and Environmental Science, 15, no. 3, 032010, 2012, IOP Publishing.

An optimization to enhance the hydraulic performances of a Francis turbine using GA and simplex optimization method with 3D CFD analysis was performed by Obrovsky and Krausová (2013). As for the parameterization of the runner, "BladeGen" was used and 36 parameters related to the geometry of the runner were considered as design variables. They used the objective function that was defined as a weighted linear combination of four performance functions related to head, efficiency, cavitation, and swirl intensity. The optimization procedure was a fully automatic cycle. GA with a population of 10 individuals was used for the first step and about 700 variants of the runner were calculated. After next 300 variants, the simplex Nelder–Mead algorithm (Nelder and Mead 1965) was used to find the final solution. Cavitation analysis of the optimized Francis turbine was performed, as shown in Figure 4.156, where the iso-surface indicates cavitation area in the runner for different cavitation coefficients.

Grafenberger et al. (2008) carried out a multi-objective optimization of a Francis turbine runner to enhance hydraulic performance using hierarchical metamodel assisted evolutionary algorithm (HMAEA) (Georgopoulou et al. 2008). They used an in-house code (3D design tool, grid generator, and CFD solver). Figure 4.157 shows design optimization procedure used in this work. The pressure coefficient and the cavitation coefficient were selected as two conflicting objective functions, and 50 parameters related to the blade profile were selected as design variables. A two-level optimization platform was set up as shown in Figure 4.158 and solved using a two-level HMAEA. On the low level, a rough exploration of the design space was performed using a Euler equation solver with coarse grid, and a more accurate and time-consuming simulation process (by solving Euler or Navier–Stokes equations with fine grid) was used on the high level focusing on the best performing region of the design space identified by the low level. The POS were found to be the multi-objective optimization results.

Multi-objective optimization of a runner to enhance the efficiency and the head of a Francis turbine was carried out by Derakhshan and Mostafavi (2011) using GA

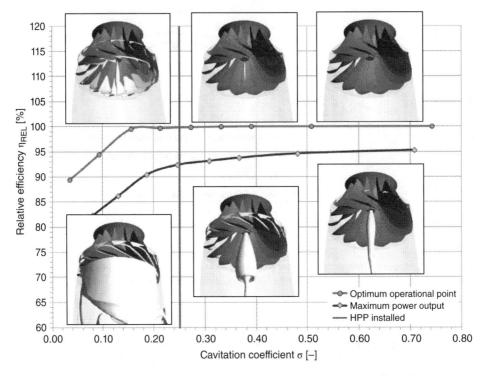

Figure 4.156 The dependency of the cavitation coefficient of the optimized runner of a Francis turbine. Source: Reproduced with permission from Obrovsky and Krausová (2013), SpringerOpen.

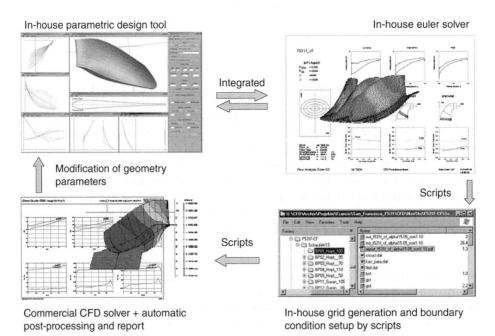

Figure 4.157 Schematic presentation of a "manual" design optimization approach using hierarchical metamodel. Source: Grafenberger et al. (2008).

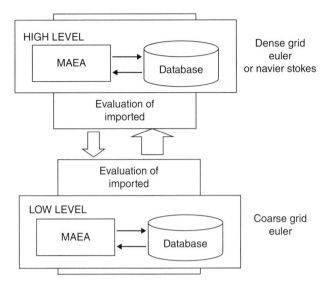

Figure 4.158 The proposed two-level optimization platform for hydro turbine blades. Source: Grafenberger et al. (2008).

(Goldberg 1989) with ANN (Hsu et al. 1995). They used CFD solver FINE™/Turbo and the grid generator AutoGrid5™ for 3D RANS analysis. The runner was composed of five sections from hub to shroud. The camber line and the blade thickness of each section were selected as design variables. The camber line was modified using a B-spline curve that was defined with five parameters and the blade thickness was modified using a fifth order Bézier curve at each section. The objective function used in their work was defined as follows:

$$F = m\left(\frac{E_t - E}{E_t}\right)^2 + n\left(\frac{\Delta P_i - \Delta P}{\Delta P_i}\right)^2, \quad K_e = \frac{m}{n} \tag{4.125}$$

where, E and ΔP are efficiency and pressure drop, respectively, $E_t = 1.0$ is the target efficiency, $\Delta P_i = -54\,000$ Pa is the initial total pressure drop, and m and n, are weighting factors. Through the optimization for $K_e = 1000$, the efficiency and the total pressure drop were improved by approximately 1.1 and 2.9%, respectively.

4.4.4.3 Draft Tubes and Others
Numerical analysis of a draft tube of the hydraulic turbine is challenging and time-consuming due to its complex flow features such as unsteadiness, turbulence, and secondary flow. Eisinger and Ruprecht (2001) were the first to perform the optimization of a draft tube. The pressure recovery factor of the draft tube was maximized through three optimization algorithms, that is EXTREM (Jacob 1982), SIMPLEX (Nelder and Mead 1965), and GA. Six cross-sectional areas along the streamline were selected as design variables. The results showed that EXTREM and SIMPLEX methods completed quickly the optimization with a proper initial guess. On the other hand, GA showed a robust behavior in optimization but required more computational time than the others. Through the optimization, the pressure recovery factor was increased by 13%.

Marjavaara and Lundstrom (2006), and Hellstrom et al. (2007) studied modifying a sharp-heeled draft tube of hydraulic turbine to a smooth heeled draft tube with various elbow radii. The elbow radius was used as the design variable during optimization. Simple gradient-based methods were employed to search for an optimal elbow radius. Marjavaara (2006) also performed a surrogate model-based optimization of a draft tube to enhance the hydraulic performance of a Francis turbine. A CFD code, ANSYS CFX-5, was used to evaluate flow physics of the hydraulic turbine. A linear combination of maximum pressure recovery factor and energy loss factor was considered to be an objective function. Through shape optimization of the draft tube, the average pressure recovery factor and the energy loss factor were improved by 0.1 and 3.4%, respectively.

Sale et al. (2009) reported an optimization method for design of stall-regulated horizontal-axis hydraulic turbine rotors. This optimization method combined a GA with blade-element momentum performance code. The optimization was performed to maximize the hydrodynamic efficiency and also to ensure that the rotor produces an ideal power curve and avoids cavitation, in terms of the chord, twist, hydrofoil distributions, and rotor speed.

Fares et al. (2011) carried out shape optimizations of diffuser and draft tube of a hydraulic turbine using GA that was developed in Java code and coupled with a commercial CFD solver in order to create a fully automated optimization algorithm. They used FLUENT 6.3.26 for the analysis of the hydraulic turbine and mesh generator, GAMBIT. In order to optimize the diffuser, the inlet curved wall of the diffuser was parametrized using a Bézier curve with 18 control points. The pressure recovery coefficient was considered as the objective function in the diffuser optimization because the pressure recovery benefits to reduce the energy loss of the hydraulic turbine. And, the pressure at the inlet was minimized in the optimization of the draft tube. Through shape optimizations using GA, the pressure recovery coefficient of the diffuser increased by 0.14%, and the static pressure at the inlet of the draft tube decreased by 1.65%.

4.4.4.4 Pump-Turbines

Yang and Xiao (2014) performed a multi-objective optimization of a pump-turbine to enhance efficiency using 3D RANS analysis. MOGA and RSA (or RSM) surrogate models with an orthogonal DoE method were used for the optimization. BladeGen was used as a 3D geometry modeler, ANSYS ICEM CFD as a grid generator, ANSYS CFX as a CFD solver, and MATLAB as a surrogate modeler. Figure 4.159 shows the whole design procedure. The optimization was performed in an automatic method by incorporating all the necessary software into the "Isight" platform. The loading distribution on the blade and the stacking condition were considered for parameterization of the blade geometry. Eight parameters of blade shape were used as design variables. The efficiencies of both pump and turbine modes at the designed mass flow rate (i.e., η_p and η_t, respectively) and the 80% designed mass flow rate (i.e., $\eta_{p,80}$ and $\eta_{t,80}$, respectively), were selected as the objective functions. Through the optimization, the Pareto-front curve was obtained as shown in Figure 4.160. The result of the final design was validated by an experimental test. The efficiencies of pump and turbine modes at 80% designed mass flow rate were increased by 1.18 and 0.24%, respectively.

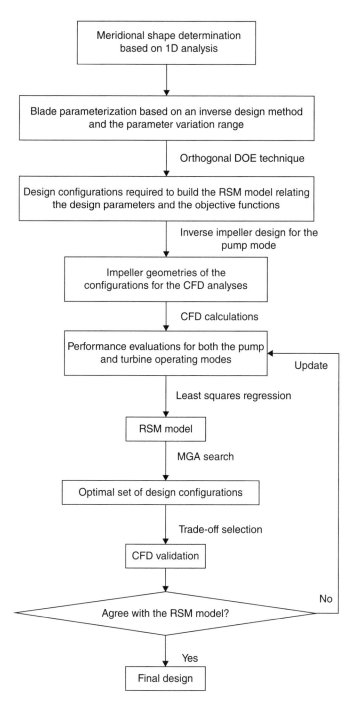

Figure 4.159 Pump-turbine design procedure. Source: Reproduced with permission from Yang and Xiao (2014), Copyright © 2014 by American Society of Mechanical Engineers.

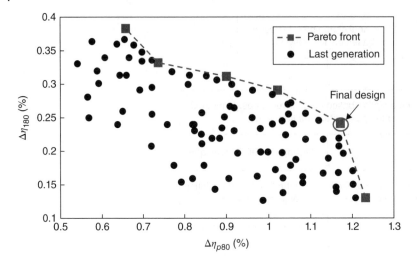

Figure 4.160 Pareto front for the optimization results of a pump-turbine. Source: Reproduced with permission from Yang and Xiao (2014), Copyright © 2014 by American Society of Mechanical Engineers.

4.5 Others

4.5.1 Regenerative Blowers

Regenerative blowers have advantages of high pressure at low flow rates, high operating stability, simple structure, and low manufacturing cost, and thus have been widely used for automotive fuel cells, air supply systems, and waste water circulators. However, due to high noise levels and low efficiency of regenerative blowers, efficiency enhancement and noise reduction are important in their design.

Badami and Mura (2010, 2012a,b) performed a series of works to investigate the aerodynamic performance of a regenerative blower used for hydrogen recirculation in a proton exchange membrane fuel cell shown in Figure 4.161. First, Badami and Mura (2010) performed a theoretical analysis using a 1D model that was slightly modified from the theoretical model proposed for regenerative pumps on the basis of momentum exchange theory by Badami (1997). The efficiency, head coefficient, and flow coefficient of the regenerative blower were derived in terms of the geometric parameters shown in Figure 4.162 under the assumptions of compressible working fluid and constant angular speed along the side channel. The performance map of the regenerative blower was obtained through the 1D theoretical model and was validated compared with experimental data. Furthermore, the same authors, Badami and Mura (2012b) investigated the leakage effects on performance of the regenerative blower through an experimental test. Two different clearances of 0.3 and 0.5 mm between the impeller and casing, were considered in their work. They estimated the efficiency and the head coefficient in a range of flow rate from zero to maximum flow rate, compared with the results of the 1D theoretical model that were obtained in their previous work (Badami and Mura 2010). The results indicated that the efficiency and the head coefficient sharply decreased with

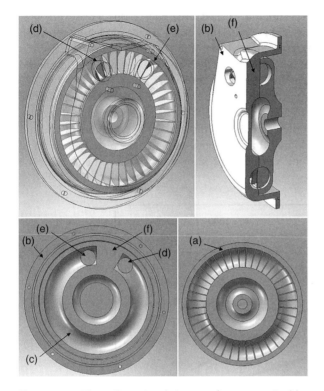

Figure 4.161 Three-dimensional pictures of a regenerative blower. Source: Reprinted from Badami and Mura (2010), with permission from Elsevier.

Figure 4.162 Main geometrical parameters of the impeller of a regenerative blower. Source: Reprinted from Badami and Mura (2010), with permission from Elsevier.

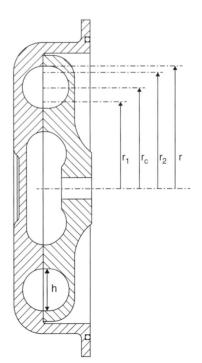

increase in the clearance. The experimental results obtained in this work were predicted well by the 1D theoretical model for both the two clearance models as shown in
Figure 4.163. In order to understand the internal flow of the regenerative blower, Badami
and Mura (2012a) also carried out a 3D RANS analysis and the results were compared
with the results of 1D theoretical analysis (Badami and Mura 2010). They used a CFD
code, ANSYS CFX for steady 3D RANS analysis of the flow through the regenerative
blower. In order to simplify the computational domain, the leakage flow was not considered in the analysis. The computational domain was discretized with tetrahedral meshes
and the number of meshes was adopted through a grid-dependency test. The results of
3D RANS analysis for head and efficiency curves, showed good agreements with the

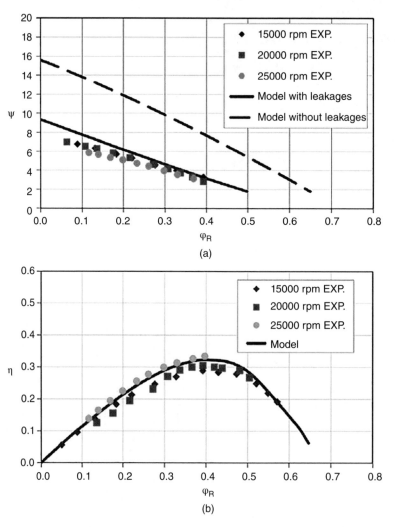

Figure 4.163 Model results versus experimental data for a clearance equal to 0.5 mm for a
regenerative blower. (a) dimensionless theoretical results for different vane head curve and (b)
efficiency curve. Source: Reprinted from Badami and Mura (2012b), with permission from Elsevier.

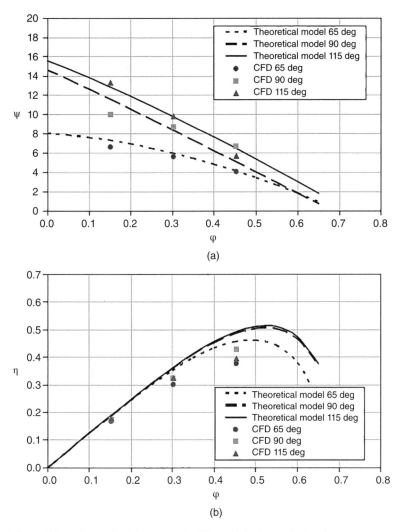

Figure 4.164 Comparison between the CFD and 1-D theoretical analysis in a regenerative blower: (a) head curves and (b) efficiency curves. Source: Reprinted from Badami and Mura (2012a), with permission from Elsevier.

results of 1D theoretical analysis, except for efficiency at a high flow rate as shown in Figure 4.164. The difference between the results of 3D RANS and 1D theoretical analysis was obvious for the radial velocity profile on the side-channel cross section. Because the 1D theoretical model was derived using the assumption of a constant angular speed along the side channel, a linear profile of radial velocity was obtained by the 1D theoretical model unlike the results of 3D RANS analysis.

Jang and Jeon (2014) performed a parametric study of a regenerative blower using 3D RANS analysis with an SST turbulence model. They used a commercial CFD code, ANSYS CFX, for steady 3D RANS analysis of the flow in the regenerative blower. The numerical results show that non-symmetric recirculation flow is found in the blade passage near the inlet and outlet (Planes 1 and 3), while symmetric recirculation is found at the middle (Plane 2), as shown in Figure 4.165. In order to find the effects of the thickness of impeller tip and bending angle of the impeller on the aerodynamic performance of the regenerative blower, numerical calculations were carried out by changing values of the two parameters shown in Figure 4.166. The parametric study could find designs where pressure was successfully increased up to 2.8% at the design flow rate and efficiency was increased up to 2.98% compared to a reference blower. The relatively uniform flow was observed at the outlet of the blower, as shown in the figure at the maximum efficiency condition, which reduces pressure loss by velocity defect.

Mekhail et al. (2015) investigated the effects of inlet blade angle on the performance of a regenerative blower using theoretical, experimental, and numerical methods. Similar to the previous research by Badami and Mura (2010), a 1D theoretical model was derived based on the momentum exchange theory. For numerical analysis, they used a commercial CFD code, ANSYS CFX 16.1, for internal flow analysis of the regenerative blower using 3D RANS equations. As for the experimental test, four different inlet blade angles, 90°, 115°, 125°, and 135°, of regenerative blowers were tested. Predictions using the 1D theoretical model showed good agreement with the 3D RANS analysis and the experimental results for the performance of the regenerative blower. The results indicated that the pressure and the efficiency of the regenerative blower depend strongly on the blade inlet angle, and the best performance was obtained at the inlet blade angle of 125°.

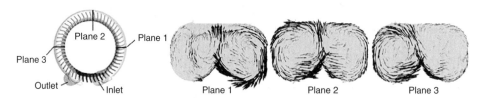

Figure 4.165 Tangential velocity vectors for reference regenerative blower. Source: Reproduced with permission from Jang and Jeon (2014).

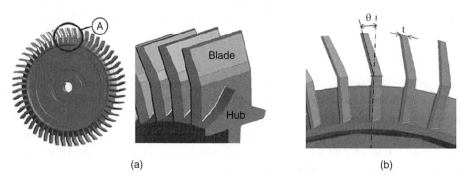

(a)   (b)

Figure 4.166 Definition of design variables (θ, t) of a regenerative blower: (a) perspective view of the impeller and (b) design variables (detailed view of a). Source: Reproduced with permission from Jang and Jeon (2014).

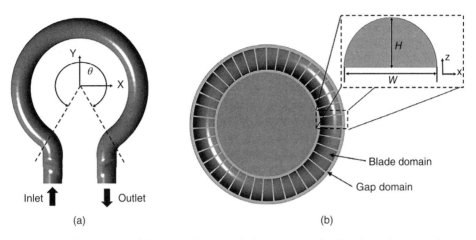

Figure 4.167 Computational domain and geometrical parameters of a side-channel regenerative blower. Source: Reproduced with permission from Heo et al. (2015b), Copyright © 2015 by American Society of Mechanical Engineers.

Heo et al. (2015b) performed both aerodynamic and aeroacoustic analyses of a regenerative blower. 3D steady and unsteady RANS equations with the SST turbulence model were used for the aerodynamic analysis, and the aeroacoustic analysis was performed using the variational formulation of the Lighthill's analogy based on the aerodynamic sources obtained from the unsteady RANS analysis. Effects of the height (H) and width (W) of the blade (or cavity), and the angle between inlet and outlet ports (θ), shown in Figure 4.167, on aerodynamic and aeroacoustic performances were investigated in this work. The results indicated that both the efficiency and overall SPL were most sensitive to H/D among the tested parameters, where D is the impeller diameter. The maximum efficiency value found in the parametric study was 37.5% at $H/D = 0.047$. The overall SPL increased as H/D and θ increased and W/D decreased. The minimum overall SPL was found to be 106.6 dB at $H/D = 0.032$.

Recently, a few optimization studies were carried out for regenerative blowers (Jang and Lee 2012; Heo et al. 2016). Jang and Lee (2012) performed an optimization of a regenerative blower using 3D RANS analysis and an RSA surrogate model. They used a commercial CFD cord, ANSYS CFX, for steady RANS calculation of the flow through a regenerative blower. Two different optimizations were performed; Cases 1 and 2 maximized the efficiency and the pressure at the outlet of the regenerative blower, respectively. The optimizations were carried out by changing the number of impeller and extension angle with constant mass flow rate. The extension angle was defined as the angle between the inlet and outlet port. Through the optimizations, the efficiency was improved up to 1.4% in Case 1, and the pressure was improved up to 3.1% in Case 2.

A multi-objective (and also multidisciplinary) optimization of a side-channel type regenerative blower was performed by Heo et al. (2016) to simultaneously improve the aerodynamic and aeroacoustic performances based on the 3D aerodynamic and aeroacoustic analyses using unsteady RANS equations and the variational formulation of Lighthill's analogy and the results of the parametric study performed by Heo et al. (2015b). The efficiency and pressure rise predicted by the preliminary design system

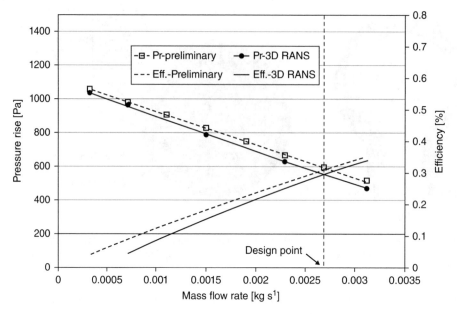

Figure 4.168 Validation of preliminary performance analysis results compared to RANS analysis results for a regenerative blower. Source: With kind permission from Springer Science+Business Media, © 2016, 1197–1208, Heo et al. (2016).

(Lee et al. 2013) showed qualitatively good agreements with the RANS analysis results, as shown in Figure 4.168. The blade height-to-impeller diameter (H/D) and blade width-to-impeller diameter (W/D) ratios and the angle between the inlet and outlet port (θ), were used as design variables (see Figure 4.167). Values of the two objective functions, that is the efficiency and SPL at the design point, were calculated at the experimental points selected by the LHS in the design space to construct RBNN surrogate models for the objective functions. POS were found using a hybrid MOEA as shown in Figure 4.169. At three Pareto-optimal designs, which were chosen arbitrarily from the POS, 0.23, 1.50, and 3.77% improvements in efficiency and 13.97, 13.38, and 5.13 dB reductions in the overall SPL were obtained compared with a reference design. To validate the performance of the optimum design, an experimental performance test for one of the arbitrarily chosen designs was performed. The predicted efficiency and overall SPL were confirmed by the experimental test results with relative errors of 7.32 and 6.52%, respectively.

4.5.2 Others

Ooi (2005) optimized the performance of a rolling piston compressor for refrigerators (Figure 4.170) using a mathematical model proposed by Ooi and Chai (1998). Six geometric parameters related to the cylinder, roller, and vane of the rolling piston compressor were defined as design variables. The objective of this work was to maximize the inverse of the total frictional power loss, which is induced by each component of the compressor. In this work, a complex optimization method of Box (1965), which is based on a direct search method, was used to find the optimal solution. The results showed

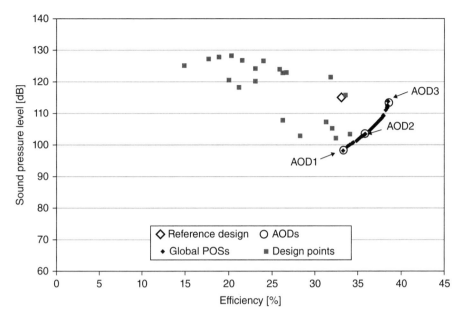

Figure 4.169 Global POS for a regenerative blower. Source: With kind permission from Springer Science+Business Media, © 2016, 1197–1208, Heo et al. (2016).

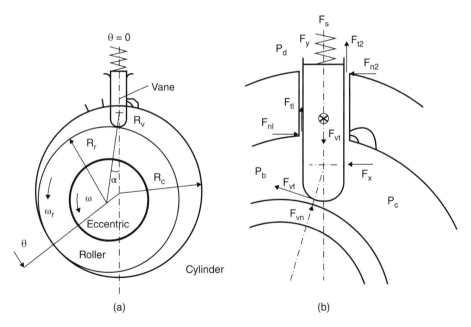

Figure 4.170 Some notations of a rolling piston compressor. (a) Some geometrical notations and (b) forces notations. Source: Reprinted from Ooi (2005), with permission from Elsevier.

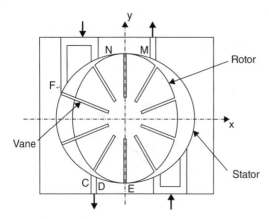

Figure 4.171 Schematic drawing of a sliding vane rotary compressor. Source: Reproduced with permission from Huang and Tsay (2009), Copyright © 2009 by American Society of Mechanical Engineers.

that through the optimization the objective function value was improved by 2.4 times and significant reduction of mechanical loss could enhance the mechanical efficiency by 14% compared to a reference design.

Casoli et al. (2008) presented a numerical analysis and an optimization of external gear pumps. An in-house code developed for simulation of gear pumps and motors was used for the numerical analysis. This optimization was particularly focused on the geometry of the recesses machined on the bushings. The objective function was defined as a linear combination of the parameters related to the volumetric efficiency, the delivery pressure ripple, the cavitation onset, and the maximum pressure peak during the meshing process. The optimum algorithm based on RSA model and steepest descent method used by Vacca and Cerutti (2007) was used in this work. Optimum design of the recesses was proposed, and a prototype of the pump with the proposed bearing blocks was tested. An experiment for volumetric efficiency and delivery pressure ripple proved the performance of the new design.

Huang and Tsay (2009) optimized the mechanical performance of a sliding vane rotary compressor (Figure 4.171) using a mathematical analysis. Design variables included dimensions of the vane, rotor rotational speed, and angular locations of the inlet and outlet ports. The mechanical efficiency considering the compression power of air and loading on the vane was employed as an objective function. The GA, which is a probability-based search method, was used to find the optimal solutions. They evaluated the effects of GA parameters such as population size, mutation rate, and crossover rate on the optimized solution, and showed that the optimized solution was significantly affected by GA parameters as shown in Figures 4.172–4.174. It was suggested that proper values of GA parameters improve the reliability of the optimized solution. Based on these analyses, a 12% increment in the mechanical efficiency of the compressor was obtained in this work compared to that obtained by Huang (1999).

Liu et al. (2010a) performed a single-objective optimization of a scroll compressor (Figure 4.175) using a mathematical model. The mechanical loss of the scroll compressor was used as an objective function, and six geometric parameters associated with the journal bearing and two geometric parameters for the thrust bearing (totally eight geometric parameters) were selected as design variables. Eight constraint conditions for lubrications and allowable strength were designated to make sure of ideal operation. Also, SQP (Arora 2004), which is a gradient-based search method, was used to find the

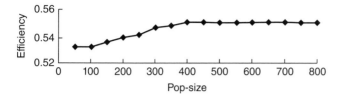

Figure 4.172 Effect of population size on mechanical efficiency. Source: Reproduced with permission from Huang and Tsay (2009), Copyright © 2009 by American Society of Mechanical Engineers.

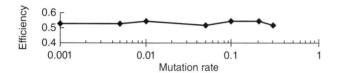

Figure 4.173 Effect of mutation rate on mechanical efficiency. Source: Reproduced with permission from Huang and Tsay (2009), Copyright © 2009 by American Society of Mechanical Engineers.

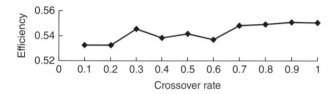

Figure 4.174 Effect of crossover rate on mechanical efficiency. Source: Reproduced with permission from Huang and Tsay (2009), Copyright © 2009 by American Society of Mechanical Engineers.

optimal solution. As a result, they found that the optimal design can reduce the mechanical loss by 15.8% compared to the reference design.

Shi et al. (2014) optimized an axial piston pump (Figure 4.176) using a design optimization technique. The objectives of the optimization were to avoid the oil shock and to make the pressure gradient steady. Dynamic pressure of piston chamber and oil shock were mathematically modeled. They validated the analysis results by performing a real pump-flow experiment. The shape of the piston pump was optimized using the particle swarm optimization (PSO) algorithm (Kennedy and Eberhart 1997) and the optimization procedure is shown in Figure 4.177. Through the optimization, the maximum pressure gradient became only 1.9 times larger than the minimum pressure gradient, which decreased by 70.8% compared with the original value, and the flow fluctuation decreased by 21.4%.

A mathematical model of the working process of a piston compressed air engine was proposed and a multi-objective optimization of the piston air engine (Figure 4.178) was performed by Yu et al. (2014) based on a mathematical model. The commercial software MATLAB® and Simulink® were used to evaluate the performance of the piston engine based on the mathematical model. Operational parameters, such as the intake pressure and valve lift, were selected as design variables based on a parametric study, and the output power and energy efficiency were set as objective functions. In this work, a concept of improved NSGA-II (Deb et al. 2002) was introduced as an optimization algorithm. It was confirmed that the results of the improved NSGA-II were superior to

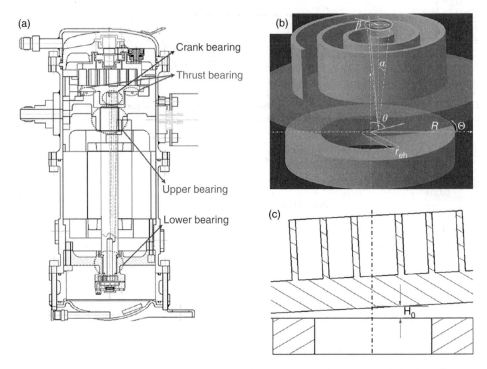

Figure 4.175 Bearing components: (a) schematic of the three journal bearings and the thrust bearing, (b) parameters of the thrust bearing, and (c) nominal height of oil-film. Source: Reprinted from, Liu et al. (2010a), with permission from Elsevier.

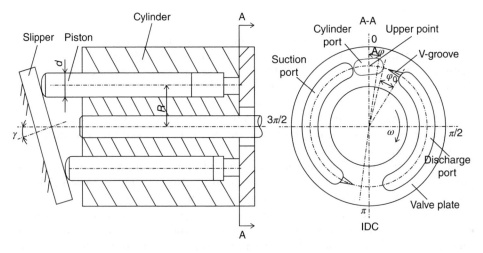

Figure 4.176 Cutaway view of a piston pump and the structure of the valve plate. Source: Reproduced with permission from Shi et al. (2014), Hindawi.

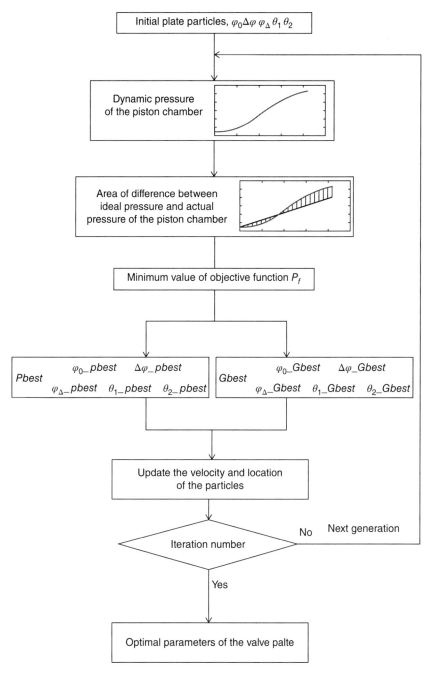

Figure 4.177 Flowchart of the PSO. Source: Reproduced with permission from Shi et al. (2014), Hindawi.

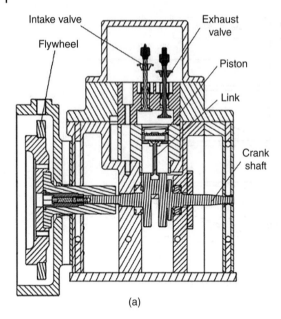

(a)

Figure 4.178 The thermodynamic analysis diagram of a compressed air engine; (a) structure of the piston CAE and (b) configuration of the piston CAE. Source: Reproduced with permission from Yu et al. (2014).

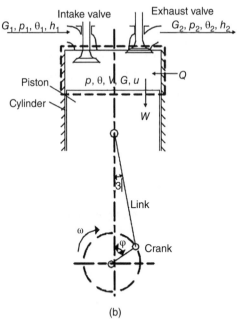

(b)

those of the NSGA-II model in proximity and diversity. As the intake pressure or the valve lift increased, the output power increased, while the energy efficiency decreased. They suggested that, since the major role in regulating intake pressure is to meet variable speeds and road conditions, appropriate adjustment of the intake valve lift could enhance energy efficiency.

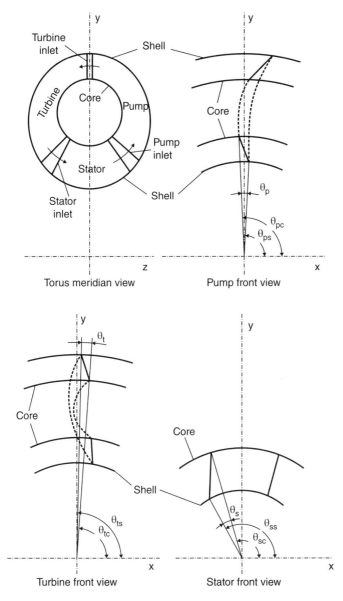

Figure 4.179 Torque converter torus and definition of inlet deflection angle. Source: Reproduced with permission from Liu et al. (2015), Copyright © 2015 by American Society of Mechanical Engineers.

Liu et al. (2015) performed a multi-objective optimization of a torque converter using 3D RANS analysis with SST turbulence model. The inlet deflection angles (θ_p, θ_t, and θ_s in Figure 4.179) related to three components (pump, turbine, and stator) of the torque converter were considered to be the design variables in this work. For the optimization, three performance parameters were defined as the objective functions;

peak efficiency, stall torque ratio, and stall pump capacity factor (Heldt 1955). To search for multi-objective optimization solutions, the archive-based micro-GA (AMGA) (Tiwari et al. 2008) was used in this work. The AMGA found Pareto optimal solutions by using a large external archive, which was based on a diversity information, with a small population to reduce the computing time.

References

AFNOR (1999). ISO 5801: Industrial fans, performance testing using standardized airways.

Ainley, D. G., and Mathieson, G. C. (1951). *A method of performance estimation for axial-flow turbines* (No. ARC-R/M-2974). Aeronautical Research Council London (UK).

Al-Zubaidy, S. N. J. (1990). *Toward Automating the Design of Centrifugal Impellers*, ASME Fluid Machinery Forum, Spring Meeting of the Fluids Engineering Division, June 4–7, Toronto, Ontario, Canada, pp. 41–47.

AMCA (1999). AMCA Standard 210-99/ASHRAE Standard 51-1999, Laboratory Methods of Testing Fans for Aerodynamic Performance Rating, AMCA and ASHRAE.

Anderson, M. (1997). Design of Experiments. *American Institute of Physics* 3 (3): 24–26.

Arad, N. and Reisfeld, D. (1995). Image warping using few anchor points and radial functions. *Computer Graphics Forum* 14 (1): 35–46. https://doi.org/10.1111/1467-8659 .1410035.

Arnone, A. (1994). Viscous analysis of three-dimensional rotor flow using a multigrid method. *ASME Journal of Turbomachinery* 116 (3): 435–445.

Arnone, A., Marconcini, M. Rubechini, F. Schneider, A., and Alba, G. (2009). Kaplan turbine performance prediction using CFD: An artificial neural network approach, *HYDRO 2009 Conference Paper* no. 263, Lyon, France.

Arora, J.S. (2004). *Introduction to Optimum Design*. New York: McGraw-Hill.

Aulich, A. L., Goerke, D., Blocher, M., Nicke, E., and Kocian, F. (2013). Multidisciplinary automated optimization strategy on a counter rotating fan, Proceedings of ASME Turbo Expo 2013: Turbine Technical Conference and Exposition, June 3–7, 2013, San Antonio, Texas, USA. GT2013-94259.

Avellan, F. (2004). Introduction to cavitation in hydraulic machinery, *The 6th International Conference on Hydraulic Machinery and Hydrodynamics,* Timisoara, Romania.

Badami, M. (1997). Theoretical and experimental analysis of traditional and new periphery pumps, *SAE Technical Paper* 971074.

Badami, M. and Mura, M. (2010). Theoretical model with experimental validation of a regenerative blower for hydrogen recirculation in a PEM fuel cell system. *Energy Conversion and Management* 51: 553–560.

Badami, M. and Mura, M. (2012a). Comparison between 3D and 1D simulations of a regenerative blower for fuel cell applications. *Energy Conversion and Management* 55: 93–100.

Badami, M. and Mura, M. (2012b). Leakage effects on the performance characteristics of a regenerative blower for the hydrogen recirculation of a PEM fuel cell. *Energy Conversion and Management* 55: 93–100.

Ballesteros, R., Velarde, S., and Santolaria, C. (2002). Turbulence intensity measurements in a forward-curved blades centrifugal fan, Proceedings of the XXIst IAHR Symposium on Hydraulic Machinery and Systems, September 9–12, Lausanne.

Ballesteros-Tajadura, R., Velarde-Suárez, S., Hurtado-Cruz, J.P., and Santolaria-Morros, C. (2006). Numerical calculation of pressure fluctuations in the volute of a centrifugal fan. *ASME Journal of Fluids Engineering* 128 (3): 359–369.

Banaszek, M. and Tesch, K. (2010). Rotor blade geometry optimization in Kaplan turbine. *TASK Quarterly* 14 (3): 209–225.

Baskharone, E.A. (2006). *Principles of Turbomachinery in Air-Breathing Engines.* Cambridge University Press 9780511616846.

Benini, E. (2004). Three-dimensional multi-objective design optimization of a transonic compressor rotor. *Journal of Propulsion and Power* 20 (3): 559–565.

Benini, E. (ed.) (2013). Advances in aerodynamic design of gas turbine compressors. In: *Progress in Gas Turbine Performance, IntechOpen.* https://doi.org/10.5772/2797, ISBN: 978-953-51-1166-5.

Benini, E. and Cenzon, M. (2009). Calibration of a meanline centrifugal pump model using evolutionary algorithms. *Proceedings of the Institution of Mechanical Engineers, Part A: Journal of Power and Energy* 223 (7): 835–847.

Benini, E. and Toffolo, A. (2001). A parametric method for optimal design of two-dimensional cascades. *Proceedings of the Institution of Mechanical Engineers, Part A: Journal of Power and Energy* 215 (4): 465–473. https://doi.org/10.1243/0957650011538721.

Benini, E. and Toffolo, A. (2002). Development of high-performance airfoils for axial flow compressors using evolutionary computation. *Journal of Propulsion and Power* 18 (3): 544–554. https://doi.org/10.2514/2.5995.

Benini, E., and Tourlidakis, A. (2001, June). Design optimization of vaned diffusers for centrifugal compressors using genetic algorithms. In *15th AIAA Computational Fluid Dynamics Conference*, p. 2583.

Benini, E., Boscolo, G., and Garavello, A. (2008, January). Assessment of loss correlations for performance prediction of low reaction gas turbine stages. In: *ASME 2008 International Mechanical Engineering Congress and Exposition*, 177–184. American Society of Mechanical Engineers.

Bitter, J. (2007). One-Dimensional Modeling of Centrifugal Flow Vaned Diffusers. MSc Thesis, Brigham Young University, Provo, UT.

Bonaiuti, D. and Pediroda, V. (2001). Aerodynamic optimization of an industrial centrifugal compressor impeller using genetic algorithms. In: *Evolutionary Methods for Design, Optimization and Control, CIMNE, Barcelona, Spain*, 467–472.

Bonaiuti, D., Arnone, A., Ermini, M., and Baldassarre, L. (2002). *Analysis and Optimization of Transonic Centrifugal Compressor Impellers Using the Design of Experiments Technique*, ASME Paper GT-2002-30619.

Box, M.J. (1965). A new method of constraint optimization and comparison with other methods. *The Computer Journal* 8: 33–41.

Braun, O., Kueny, J. L., and Avellan, F. (2005). Numerical analysis of flow phenomena related to the unstable energy-discharge characteristic of a pump-turbine in pump mode, *ASME Fluids Engineering Division Summer Meeting and Exhibition*, Huston, Texas.

Cai, N., and Xu, J. (2001). Aerodynamic-aeroacoustic performance of parametric effects for skewed-swept rotor, Proceedings of ASME Turbo Expo 2001, June 4–7, New Orleans, Louisiana, USA, 2001-GT-0354.

Cai, N., Xu, J., and Benaissa, A. (2003). Aerodynamic and aeroacoustic performance of a skewed rotor, Proceedings of ASME Turbo Expo, GT-2003-38592, June 16-19, 2003, Atlanta, Georgia, USA.

Casey, M.V. (1983). A computational geometry for the blades and internal flow channels of centrifugal compressors. *ASME Journal for Engineering for Power* 105 (2): 288–295.

Casoli, P., Vacca, A., and Berta, G.L. (2008). Optimization of relevant design parameters of external gear pumps. In: *Proceedings of the 7ᵗʰ JFPS International Symposium on Fluid Power, September 15–18*, 277–282.

Checcucci, M., Sazzini, F., Marconcini, M. et al. (2011). Assessment of a neural-network-based optimization tool: a low specific-speed impeller application. *International Journal of Rotating Machinery* 817547.

Chen, X., Kim, K.Y., and Kim, S.Y. (1996). Numerical simulation on three-dimensional viscous flow in a multiblade centrifugal fan. In: *ASME FED Conference*, vol. 238, 647–652.

Chen, L., Sun, F., and Wu, F.C.W. (2005). Optimum design of a subsonic axial-flow compressor stage. *Applied Energy* 80 (2): 187–195.

Chen, S., Wang, D., and Sun, S. (2011). Bionic fan optimization based on taguchi method. *Engineering Applications of Computational Fluid Mechanics* 5 (3): 302–314.

Cherny, S.G., Sharov, S.V., Skorospelov, V.A., and Turuk, P.A. (2003). Methods for three-dimensional flows computation in hydraulic turbines. *Russian Journal of Numerical Analysis Mathematical Modeling* 18 (2): 87–104.

Cherny, S.G., Chirkov, D.V., Lapin, V.N. et al. (2005). 3D Euler flow simulation in hydro turbines: unsteady analysis and automatic design. In: *Notes on Numerical Fluid Mechanics and Multidisciplinary Design*, vol. 93, 33–51. Heidelberg: Springer.

Chirkov, D., Avdyushenok, A., Panov, L. et al. (2012). CFD simulation of pressure and discharge surge in Francis turbine at off-design conditions. *IOP Conference Series: Earth and Environmental Science* 15 (3): https://doi.org/10.1088/1755-1315/15/3/032038.

Choi, J.H., Kim, K.Y., and Chung, D.S. (1997). *Numerical Optimization for Design of an Automobile Cooling Fan*, 73–77. SAE International.

Chung, K.N., Kim, Y.I., Sung, J.H. et al. (2005). A study of optimization of blade section shape for a steam turbine. In: *ASME 2005 Fluids Engineering Division Summer Meeting*, 53–57. American Society of Mechanical Engineers.

Chunxi, L., Ling, W.S., and Yakui, J. (2011). The performance of a centrifugal fan with enlarged impeller. *Energy Conversion and Management* 52: 2902–2910.

Collette, Y. and Siarry, P. (2003). *Multiobjective Optimization: Principles and Case Studies*. New York: Springer.

Corsini, A. and Rispoli, F. (2005). Flow analyses in a high-pressure axial ventilation fan with a non-linear eddy-viscosity closure. *International Journal of Heat and Fluid Flow* 26: 349–361.

Cosentino, R., Alsalihi, Z., and Van den Braembussche, R.A. (2001). Expert system for radial impeller optimization. In: *Proceedings of the Fourth European Conference on Turbomachinery*, 481–490.

Craig, H.R.M. and Cox, H.J.A. (1970). Performance estimation of axial flow turbines. *Proceedings of the Institution of Mechanical Engineers* 185 (1): 407–424.

Cumpsty, N.A. (1989). *Compressor Aerodynamics*. Longman.

Deb, K., Pratap, A., Agarwal, S., and Meyarivan, T. (2002). A fast and elitist multiobjective genetic algorithm: NSGA-II. *IEEE Transactions on Evolutionary Computation* 6 (2): 182–197.

Denton, J.D. (1978). Throughflow calculations for transonic axial flow turbines. *Journal of Engineering for Power* 100 (2): 212. https://doi.org/10.1115/1.3446336.

Derakhshan, S. and Mostafavi, A. (2011). Optimization of GAMM Francis turbine runner. *International Journal of Mechanical, Aerospace, Industrial, Mechatronic and Manufacturing Engineering* 5 (11): 2139–2145.

Dixon, S.L. and Hall, C.A. (2014). *Fluid Mechanics and Thermodynamics of Turbomachinery*. Amsterdam: Butterworth-Heinemann/Elsevier.

Downie, R.J., Thompson, M.C., and Wallis, R.A. (1993). An engineering approach to blade designs for low to medium pressure rise rotor-only axial fans. *Experimental Thermal Fluid Science* 6: 376–401.

Drela, M. (1986). *Two-Dimensional Transonic Aerodynamic Design and Analysis Using the Euler Equations*. Cambridge, MA: Gas Turbine Laboratory, Massachusetts Institute of Technology.

Drela, M., and Youngren, H. (1995). A user's guide to MISES 2.3. MIT Laboratory Computational Aerospace Science Laboratory report.

Drela, M., and Youngren, H. (2008). A user's guide to MISES 2.63. MIT Laboratory Computational Aerospace Science Laboratory report.

Drtina, P. and Sallaberger, M. (1999). Hydraulic turbines – basic principles and state-of-theart computational fluid dynamics applications. *Proceedings of the Institution of Mechanical Engineers: Part C* 213 (1): 85–102.

Dunham, J. (1997). Modelling of spanwise mixing in compressor through-flow computations. *Proceedings of the Institution of Mechanical Engineers, Part A: Journal of Power and Energy* 211 (3): 243–251.

Eck, B. (1973). *Fans*. Oxford, UK: Pergamon Press.

Eckardt, D. (1975). Instantaneous measurements in the jet-wake discharge flow of a centrifugal compressor impeller. *ASME Journal of Engineering for Power* 3: 337–346.

Eckardt, D. (1979). Flow field analysis of radial and backswept centrifugal compressor impellers. I-Flow measurements using a laser velocimeter. In: *Performance Prediction of Centrifugal Pumps and Compressors*, 77–86. New York: ASME.

Eckardt, D. (1987). *Centrifugal Compressor Data Book*. Munich, Germany: MTU.

Eckardt, D., and Trültzsch, B. R. (1977). Vergleichende Stromungsuntersuchungen an Drei Radialverdichter-Laufra ¨dern mit Konventio-nellen Messverfahren, FVV Research Report (Forschungsberichte) #237.

Eisinger, R. and Ruprecht, A. (2001). Automatic shape optimization of hydro turbine components based on CFD. *TASK Quarterly* 6: 101–111.

Enomoto, Y., Kurosawa, S., and Kawajiri, H. (2012). Design optimization of a high specific speed Francis turbine runner, *Proceedings of the 26th IAHR Symposium on Hydraulic Machinery and Systems, 19–23 August, Bejing, China*.

Envia, A. and Kerschen, E.J. (1986). Noise generated by convected gusts interacting with swept airfoil cascades. *AIAA Journal* 86, *Proceedings of AIAA 10th Aeroacoustics Conference, July 9-11, Seattle, Washington*, AIAA-86-1872.

Estevadeordal, J., Gogineni, S., Goss, L. et al. (2000). Study of flow field interactions in a transonic compressor using DPIV. In: *Proceedings of the 38th Aerospace Sciences Meeting and Exhibit, Reno, NV*, 10–13. (AIAA Paper 00-03).

Fares, R., Chen, X., and Agarwal, R. (2011). Shape optimization of an axisymmetric diffuser and a 3D hydro-turbine draft tube using a genetic algorithm, *49th AIAA Aerospace Sciences Meeting including the New Horizons Forum and Aerospace Exposition, Orlando, Florida*.

Farlow, S.J. (1984). *Self-Organizing Method in Modeling: GMDH Type Algorithm*. Marcel Dekker Inc.

Fedala, D., Koudri, S., Rey, R. et al. (2006). Incident turbulence interaction noise from an axial fan. In: *Collection of Technical Papers, 12th AIAA/CEAS Aeroacoustics Conference*, vol. 2, 1003–1013.

Fletcher, R. (1987). *Practical Methods of Optimization*. Chichester: Wiley https://doi.org/ 10.1097/00000539-200101000-00069.

Fottner, L. (1990). *Test Cases for Computation of Internal Flows in Aero Engine Components. (Propulsion and Energetics Panel Working Group)(Exemples de Tests pour le Calcul des Ecoulements Internes dans les Organes des Moteurs d'Avion)* (No. AGARD-AR-275). Advisory Group For Aerospace Research and Development Neuilly-Sur-Seine (France).

Francis, J.B. (1909). *Lowell Hydraulic Experiments, 5th Edition, Hydraulic Losses in the Spiral Casing of a Francis Turbine*. Princeton, NJ: Van Nostrand.

Fukano, T., Kodama, Y., and Senoo, Y. (1977). Noise generated by low pressure axial flow fans. I: Modeling of the turbulent noise. *Journal of Sound and Vibration* 50: 63–74.

Genter, C., Sallaberger, M., Widmer, C., Braun, O., and Staubli, T. (2012). Numerical and experimental analysis of instability phenomena in pump turbines, *26th IAHR Symposium on Hydraulic Machinery and Systems, Beijing, China*.

Georgopoulou, H., Kyriacou, S., Giannakoglou, K., Grafenberger, P., and Parkinson, E. (2008). Constrained multi-objective design optimization of hydraulic components using a hierarchical metamodel assisted evolutionary algorithm. Part 1: Theory, *24th IAHR Symposium on Hydraulic Machinery and Systems, Foz do Iguassu, Brazil*.

Giles, M.B. and Drela, M. (1987). Two-dimensional transonic aerodynamic design method. *AIAA Journal* 25 (9): 1199–1206. https://doi.org/10.2514/3.9768.

Göde, E., and Ryhming, I. L. (1987). 3D-Computation of the flow in a Francis runner. Sulzer Technical Review No. 4.

Göde, E., Cuénod, R., and Pestalozzi, J. (1989). Visualization of flow phenomena in a hydraulic turbine based on 3D flow computations, *Proceedings of the Waterpower '89, August 23–25, Niagara Falls, New York, USA*.

Goel, T., Haftka, R.T., Shyy, W., and Queipo, N.V. (2007). Ensemble of surrogates. *Structural and Multidisciplinary Optimization* 33 (3): 199–216.

Goldberg, D.E. (1989). *Genetic Algorithms in Search, Optimization and Machine Learning*. Addison-Wesley https://doi.org/10.5860/CHOICE.27-0936.

Gomes, J. (1999). *Warping and Morphing of Graphical Objects*, vol. 1. Morgan Kaufmann.

González, J., Fernández, J., Blanco, E., and Santolaria, C. (2002). Numerical simulation of the dynamic effects due to impeller-volute interaction in a centrifugal pump. *ASME Journal of Fluids Engineering* 124: 348–355.

Grafenberger, P., Parkinson, E., Georgopoulou, H., Kyriacou, S., and Giannakoglou, K. (2008). Constrained multi-objective design optimization of hydraulic components using a hierarchical metamodel assisted evolutionary algorithm. Part 2: Applications, *24th IAHR Symposium on Hydraulic Machinery and Systems*, Foz do Iguassu, Brazil.

Grekula, M., and Bark, G. (2001). Experimental study of cavitation in a Kaplan model turbine, *CAV2001*: session B9.004.

Gu, C. (1984). Theory and applications of finite element approximate solution method (FEASM). *ACTA Mechanica Sinica* 16 (6): 1–11.

Gui, L., Gu, C., and Chang, H. (1989). Influence of Splitter Blades on the Centrifugal Fan Performances, ASME Turbo Expo. 89, GT-33.

Guo, E.M. and Kim, K.Y. (2004). Three-dimensional flow analysis and improvement of slip factor model for forward-curved blades centrifugal fan. *KSME International Journal* 18 (2): 302–312.

Han, S.Y. and Maeng, J.S. (2013). Shape optimization of cutoff in a multi-blade fan/scroll system using neural network. *International Journal of Heat and Mass Transfer* 46: 2833–2839.

Han, S. Y., Maeng J. S., and Yoo, D. H. (2003). Shape optimization of cutoff in a multiblade fan/scroll system using response surface methodology. Numerical Heat Transfer, B. 43, 87–98.

Hasmatuchi, V., Farhat, M., Maruzewski, P., and Avellan, F. (2009). Experimental investigation of a pump-turbine at off-design operation conditions, *Proceedings of the 3rd International Meeting of the Workgroup on Cavitation and Dynamic Problems in Hydraulic Machinery and Systems*, Brno, Czech Republic.

Heldt, P.M. (1955). *Torque Converters or Transmissions*. New York: Childon Company.

Hellstrom, J., Marjavaara, B., and Lundstrom, T. (2007). Parallel CFD simulations of an original and redesigned hydraulic turbine draft tube. *Advances in Engineering Software* 38 (5): 338–344.

Heo, M.W., Kim, J.H., and Kim, K.Y. (2015a). Design optimization of a centrifugal fan with splitter blades. *International Journal of Turbo and Jet Engines* 32 (2): 143–154.

Heo, M. W., Seo, T. W., Lee, C. S., and Kim, K. Y. (2015b). Aerodynamic and aeroacoustic analyses of a regenerative blower, Proceedings of ASME Turbo Expo 2015, *June 15–19, 2015, Montréal, Canada*, GT2015-42050.

Heo, M.W., Seo, T.W., Shim, H.S., and Kim, K.Y. (2016). Optimization of a regenerative blower to enhance aerodynamic and aeroacoustic performance. *Journal of Mechanical Science and Technology* 30 (3): 1197–1208.

Herrig, J.L., Emery, J.C., and Erwin, J.R. (1957). Systematic two-dimensional cascade tests of Naca 65-series compressor blades at low speeds. In: *NACA Technical Note*, 226. https://doi.org/10.1063/1.1715007.

Horlock, J.H. (1966). *Axial Flow Turbines*. Krieger Publishing Company.

Howard, M.A. and Gallimore, S.J. (1993). Viscous throughflow modeling for multistage compressor design. *Journal of Turbomachinery* 115 (2): 296–304.

Hsu, K., Gupta, H.V., and Sorroshian, S. (1995). Artificial neural network modeling of the rainfall-runoff process. *Water Resources Research* 31 (10): 2517–2530.

Huang, Y.M. (1999). The performance and fluid properties of a rotary compressor. *ASME Journal of Pressure Vessel Technology* 396: 99–104.

Huang, Y.M. and Tsay, S.N. (2009). Mechanical efficiency optimization of a sliding vane rotary compressor. *ASME Journal of Pressure Vessel Technology* 131 (4), Article ID: 061601): 8.

Huang, J., Corke, T.C., and Thomas, F.O. (2006). Plasma actuators for separation control of low-pressure turbine blades. *AIAA Journal* 44 (1): 51–57.

Hurault, J., Kouidri, S., Bakir, F., and Rey, R. (2010). Experimental and numerical study of the sweep effect on three-dimensional flow downstream of axial flow fans. *Flow Measurement and Instrumentation* 21: 155–165.

Iliescu, M. S., Ciocan, G. D., and Avellan, F. (2003). Two phase PIV measurements at the runner outlet in a Francis turbine, Proceedings of FEDSM'03, 4TH ASME_JSME Joint Fluids Engineering Conference, Honolulu, Hawaii, USA.

Iliescu, M.S., Ciocan, G.D., and Avellan, F. (2008). Analysis of the cavitating draft tube vortex in a Francis turbine using particle image velocimetry measurements in two-phase flow. *ASME Journal of Fluids Engineering* 130: 021105.

Jacob, H.G. (1982). *Rechnergestützte Optimierung statischer und dynamischer Systeme.* Berlin: Springer-Verlag.

Jang, C.M. and Jeon, H.J. (2014). Performance enhancement of 20kW regenerative blower using design parameters. *International Journal of Fluid Machinery and Systems* 7 (3): 86–93.

Jang, C.M. and Lee, J.S. (2012). Shape optimization of a regenerative blower used for building fuel cell system. *Open Journal of Fluid Dynamics* 2: 208–214.

Jang, C.M., Sato, D., and Fukano, T. (2005). Experimental analysis on tip leakage and wake flow in an axial flow fan according to flow rates. *ASME Journal of Fluids Engineering* 127 (2): 322–329.

Jang, C.M., Choi, S.M., and Kim, K.Y. (2008). Effects of inflow distortion due to hub Cap's shape on the performance of axial flow fan. *Journal of Fluid Science and Technology* 3 (5): 598–609.

Japikse, D. (1996). *Centrifugal Compressor Design and Performance.* White River Junction, VT: Concepts NREC.

Japikse, D. (2000). Decisive factors in advanced centrifugal compressor design and development. In: *Proceedings of the International Mechanical Engineering Congress & Exposition (IMechE)*, 1–14.

Japikse, D. (2001). *Enhanced TEIS and Secondary Flow Modeling for Diverse Compressors.* White River Junction, VT: Concepts NREC.

Japikse, D. and Baines, N.C. (1997). *Introduction to Turbomachinery.* Norwich, VT: Concepts ETI.

Japikse, D., Marscher, W.D., and Furst, R.B. (2006). *Centrifugal Pump Design and Performance.* White River Junction, VT: Concepts NREC.

Jeon, W. H., and Lee, D. J. (1997). An analysis of the flow and sound source of an annular type centrifugal fan, Fifth International Congress on Sound and Vibration. *December 15–18, Adelaide, South Australia.*

Karanth, K.V. and Sharma, N.Y. (2009). CFD analysis on the effect of radial gap on impeller-diffuser flow interaction as well as on the flow characteristics of a centrifugal fan. *International Journal of Rotating Machinery.* Article ID: 293508 8.

Keck, H. and Sick, M. (2008). Thirty years of numerical flow simulation in hydraulic turbomachines. *Acta Mechanica* 201 (1): 211–229.

Keck, H., Drtina, P., and Sick, M. (1996). Numerical hill chart prediction by means of CFD stage simulation for a complete francis turbine, *Proceedings of the XVIII IAHR Symposium, Valencia, Spain.*

Kennedy, J. and Eberhart, R.C. (1997). Discrete binary version of the particle swarm algorithm. In: *Proceedings of the IEEE International Conference on Systems, Man, and Cybernetics*, 4104–4108. FL, USA: Orlando.

Kergourlay, G., Kouidri, S., Rankin, G.W., and Rey, R. (2006). Experimental investigation of the 3D unsteady flow field downstream of axial fans. *Flow Measurement and Instrumentation* 17: 303–314.

Khalkhali, A., Farajpoor, M., and Safikhani, H. (2011). Modeling and multi-objective optimization of forward-curved blade centrifugal fans using CFD and neural networks. *Transactions of the Canadian Society for Mechanical Engineering* 35 (1): 63–79.

Khelladi, S., Kouidri, S., Bakir, F., and Rey, R. (2005). Flow study in the impeller–diffuser interface of a vaned centrifugal fan. *ASME Journal of Fluids Engineering* 127 (5): 495–502.

Kim, J.K. and Kang, S.H. (1997). *Effects of the Scroll on the Flow Field of a Sirroco Fan*, 1318–1327. Hawaii: ISROMAC-7.

Kim, K.Y. and Seo, S.J. (2004). Shape optimization of forward-curved-blade centrifugal fan with Navier–Stokes analysis. *ASME Journal of Fluids Engineering* 126: 735–742.

Kim, K.Y. and Seo, S.J. (2006). Application of numerical optimization technique to design of forward-curved blades centrifugal fan. *JSME International Journal: Series B* 49 (1): 152–158.

Kim, J.H., Choi, J.H., Husain, A., and Kim, K.Y. (2010). Performance enhancement of axial fan blade through multi-objective optimization techniques. *Journal of Mechanical Science and Technology* 24 (10): 2059–2066.

Kim, J.H., Kim, J.W., and Kim, K.Y. (2011). Axial-flow ventilation fan design through multi-objective optimization to enhance aerodynamic performance. *ASME Journal of Fluids Engineering* 133: 101101, 12.

Kim, J.H., Cha, K.H., Kim, K.Y., and Jang, C.M. (2012a). Numerical investigation on aerodynamic performance of a centrifugal fan with splitter blades. *International Journal of Fluid Machinery and Systems* 5 (4): 168–173.

Kim, J.H., Kim, J.H., Kim, K.Y. et al. (2012b). High-efficiency design of a tunnel ventilation jet fan through numerical optimization techniques. *Journal of Mechanical Science and Technology* 26 (6): 1793–1800.

Kim, J.H., Cha, K.H., and Kim, K.Y. (2013). Parametric study on a forward-curved blades centrifugal fan with an impeller separated by an annular plate. *Journal of Mechanical Science and Technology* 27 (6): 1589–1595.

Kim, J.H., Ovgor, B., Cha, K.H. et al. (2014). Optimization of the aerodynamic and aeroacoustic performance of an axial-flow fan. *AIAA Journal* 52 (9): 2032–2043.

Korpela, S.A. (2012). *Principles of Turbomachinery*. Wiley.

Kouidri, S., Fedala, D., Belamri, T., and Rey, R. (2005). Comparative study of the aeroacoustic behavior of three axial flow fans with different sweeps. Proceedings of the ASME FEDSM '05, *Huston, TX, USA*.

Kubo, T. and Murata, S. (1976). Unsteady flow phenomena in centrifugal fans. *Bulletin of the JSME* 19 (135): 1039–1046.

Kumar, P. and Saini, R.P. (2010). Study of cavitation in hydro turbines – a review. *Renewable and Sustainable Energy Reviews* 14 (1): 374–383.

Kurokawa, J., and Kitahora, T. (1994). Accurate determination of volumetric and mechanical efficiencies and leakage behavior of Francis turbine and Francis pump turbine, XVII IAHR Symposium, *Beijing, China*.

Kurokawa, J. and Sakuma, M. (1988). Flow in a narrow gap along an enclosed rotating disk with through-flow. *JSME International Journal* 31 (2): 243–251.

Kurokawa, J. and Toyoura, T. (1976). Axial thrust, disk friction torque and leakage loss of radial flow turbomachinery. In: *Proceedings of Pumps and Turbines Conference*, vol. 1. Glasgow, UK.

Kurokawa, J., Toyoukura, T., Shinjo, M., and Matsuo, K. (1978). Roughness effects on the flow along an enclosed rotating disk. *Bulletin of JSME* 21 (2): 1725–1732.

Lampart, P. (2004a). Numerical optimization of a high pressure steam turbine stage. *Journal of Computational and Applied Mechanics* 5 (2): 311–321.

Lampart, P. (2004b). Numerical optimisation of a high pressure steam turbine stage. In: *Modelling Fluid Flow*, 323–334. Berlin, Heidelberg: Springer.

Lee, S.-Y. and Kim, K.Y. (2000). Design optimization of axial flow compressor blades with three-dimensional Navier–Stokes Solver. *KSME International Journal* 14 (9): 1005–1012.

Lee, K.S., Kim, K.Y., and Samad, A. (2008). Design optimization of low-speed axial flow fan blade with three-dimensional RANS analysis. *Journal of Mechanical Science and Technology* 22: 1864–1869.

Lee, Y.T., Ahuja, V., Hosangadi, A. et al. (2011). Impeller design of a centrifugal fan with blade optimization. *International Journal of Rotating Machinery* 2011 Article ID: 537824: 16.

Lee, C., Kil, H.G., Kim, G.C. et al. (2013). Aero-acoustic performance analysis method of regenerative blower. *Journal of Fluid Machinery* 16 (2): 15–20. (in Korean).

Li, H.M. (2009). Fluid flow analysis of a single-stage centrifugal fan with a ported diffuser. *Engineering Applications of Computational Fluid Mechanics* 3 (2): 147–163.

Lipej, A. and Poloni, C. (2000). Design of Kaplan runner using multiobjective genetic algorithm optimization. *Journal of Hydraulic Research* 38: 73–79.

Liu, X., Dang, Q., and Xi, G. (2008). Performance improvement of centrifugal fan by using CFD. *Engineering Applications of Computational Fluid Mechanics* 2 (2): 130–140.

Liu, Y., Hung, C., and Chang, Y. (2010a). Design optimization of scroll compressor applied for frictional losses evaluation. *International Journal of Refrigeration* 33: 615–624.

Liu, S.H., Huang, R.F., and Lin, C.A. (2010b). Computational and experimental investigations of performance curve of an axial flow fan using downstream flow resistance method. *Experimental Thermal and Fluid Science* 34: 827–837.

Liu, C., Untaroiu, A., Wood, H.G. et al. (2015). Parametric analysis and optimization of inlet deflection angle in torque converters. *ASME Journal of Fluids Engineering* 137 (1), Article ID: 031101): 10.

Lootsma, F.A. (1972). *Numerical Methods for Nonlinear Optimization*, 69–97. New York: Academic Press.

Lotfi, O., Teixeira, J. A., Ivey, P. C., Kinghorn, I. R., and Sheard, A. G. (2006). Shape optimisation of axial fan blades using genetic algorithms and a 3D Navier–Stokes solver, Proceedings of ASME Turbo Expo 2006: Power for Land, Sea and Air, *8–11 May, Barcelona, Spain*, GT2006-90659.

Lu, F.A., Qi, D.T., Wang, X.J. et al. (2012). A numerical optimization on the vibroacoustics of a centrifugal fan volute. *Journal of Sound and Vibration* 331: 2365–2385.

Lyutov, A.E., Chirkov, D.V., Skorospelov, V.A. et al. (2015). Coupled multipoint shape optimization of runner and draft tube of hydraulic turbines. *ASME Journal of Fluids Engineering* 137: 111302.

Marjavaara, B. D. (2006). CFD Driven Optimization of Hydraulic Turbine Draft tubes using Surrogate Models, PhD thesis. Luleå University of Technology, Sweden.

Marjavaara, B.D. and Lundstrom, T. (2006). Redesign of a sharp heel draft tube by a validated CFD optimization. *International Journal for Numerical Methods in Fluids* 50 (8): 911–924.

Marsh, H. (1968). *A digital computer program for the through-flow fluid mechanics in an arbitrary turbomachine, using a matrix method, ARC, R&M, 3509*. London: HMSO/Ministry of Technology.

Massardo, A.S., Satta, A., and Marini, M. (1990). Axial flow compressor design optimization: part II – throughflow analysis. *Journal of Turbomachinery* 112 (3): 405–410.

McKay, M.D., Beckman, R.J., and Conover, W.J. (1979). A comparison of three methods for selecting values of input variables in the analysis of output from a computer code. *American Statistical Association* 21 (2): 239–245. https://doi.org/10.2307/1268522.

Mekhail, T.A.M., Dahab, O.M., Sadik, M.F. et al. (2015). Theoretical, experimental and numerical investigations of the effect of inlet blade angle on the performance of regenerative blowers. *Open Journal of Fluid Dynamics* 5: 224–237.

Mengistu, T. and Ghaly, W. (2008). Aerodynamic optimization of turbomachinery blades using evolutionary methods and ANN-based surrogate models. *Optimization and Engineering* 9 (3): 239–255. https://doi.org/10.1007/s11081-007-9031-1.

Menter, F.R. (1992). Improved two-equation k-omega turbulence models for aerodynamic flows. *NASA Technical Memorandum* 103978: 1–31. https://doi.org/10.2514/6.1993-2906.

Menter, F.R. (1994). Two-equation eddy-viscosity turbulence models for engineering applications. *AIAA Journal* 32 (8): 1598–1605. https://doi.org/10.2514/3.12149.

Menter, F.R., Kuntz, M., and Langtry, R. (2003). Ten years of industrial experience with the SST turbulence model. *Turbulence Heat and Mass Transfer* 4 (4): 625–632. https://doi.org/10.4028/www.scientific.net/AMR.576.60.

Menter, F.R., Langtry, R.B., Likki, S.R. et al. (2006). A correlation-based transition model using local variables – part I: model formulation. *Journal of Turbomachinery* 128 (3): 413. https://doi.org/10.1115/1.2184352.

Meyer, C.J. and Kröger, D.G. (2001). Numerical simulation of the flow field in the vicinity of an axial flow fan. *International Journal for Numerical Methods in Fluids* 36: 947–969.

Montgomery, D.C. (2012). Design and analysis of experiments. *Design* 2: https://doi.org/10.1198/tech.2006.s372.

Mortier, P. (1893). Fan or blowing apparatus. US Pat. No. 507,445

Mortenson, M.E. (1997). *Geometric Modeling*, 2nd Edition. Wiley.

Muntean, S., Balint, D., Susan-Resiga, R., Anton, I., and Darzan, C. (2004). 3D flow analysis in the spiral case and distributor of a Kaplan turbine, Proceedings of the 22nd IAHR Symposium, *Stockholm, Sweden*.

Murthy, K.N.S. and Lakshminarayana, B. (1986). Laser Doppler velocimeter measurement in the tip region of a compressor rotor. *AIAA Journal* 24 (5): 807–814.

Myers, R.H. and Montgomery, D.C. (1995). *Response Surface Methodology: Process and Product Optimization Using Designed Experiments*. New York: Wiley.

Nakamura, K. and Kurosawa, S. (2009). Design optimization of a high specific speed Francis turbine using multi-objective genetic algorithm. *International Journal of Fluid Machinery and Systems* 2 (2): 102–109.

Nelder, J.A. and Mead, R. (1965). A simplex method for function minimization. *Computer Journal* 7: 308–313.

Nilsson, H., and Davidson, L. (2000). A numerical comparison of four operating conditions in a Kaplan water turbine, focusing on tip clearance flow, Proceedings of the 20th IAHR Symposium on Hydraulic Machinery and Systems. Charlotte, USA.

Novak, R.A. (1967). Streamline curvature computing procedures for fluid- flow problems. *ASME Journal of Engineering for Power* 89: 478–490.

Novak, R.A. and Hearsey, R.M. (1977). A nearly three-dimensional intrablade computing system for turbomachinery. *Journal of Fluids Engineering* 99 (1): 154–166.

Obrovsky, J. and Krausová, H. (2013). Development of high specific speed Francis turbine for low head HPP. *Engineering Mechanics* 20 (2): 139–148.

Oksuz, O., Akmandor, I.S., and Kavsaoglu, M.S. (2002). Aerodynamic optimization of turbomachinery cascades using Euler/boundary-layer coupled genetic algorithms. *Journal of Propulsion and Power* 18 (3): 652–657.

Ooi, K.T. (2005). Design optimization of a rolling piston compressor for refrigerators. *Applied Thermal Engineering* 25: 813–829.

Ooi, K.T. and Chai, G.B. (1998). An analytical model for a vane spring design. *International Journal of Computer Applications in Technology* 11 (1/2): 98–108.

Oro, J.M.F., Díaz, K.M.A., Morros, C.S., and Marigorta, E.B. (2007). Unsteady flow and wake transport in a low-speed axial fan with inlet guide vanes. *ASME Journal of Fluids Engineering* 129: 1015–1029.

Osborne, W.C. (1973). *Fans*. Oxford, UK: Pergamon Press.

Oyama, A. and Liou, M.S. (2002a). Multiobjective optimization of rocket engine pumps using evolutionary algorithm. *Journal of Propulsion and Power* 18 (3): 528–535.

Oyama, A., and Liou, M. S. (2002b), Multiobjective Optimization of a Multi-Stage Compressor Using Evolutionary Algorithm, AIAA paper 2002-3535.

Pawlak, Z. (1982). Rough sets. *International Journal of Computer and Information Sciences* 11 (5): 341–356.

Pellegrini, A. and Benini, E. (2013). Multi-objective optimization of a steam turbine stage. *World Academy of Science, Engineering and Technology, International Journal of Mechanical, Aerospace, Industrial, Mechatronic and Manufacturing Engineering* 7 (7): 1514–1527.

Pelton, R. J. (2007). One-Dimensional Radial Flow Turbomachinery Performance Modeling. MSc Thesis, Brigham Young University, Provo, UT.

Peng, G., Fujikawa, S., Cao, S., and Lin, R. (1998a). An advanced three-dimensional inverse model for the design of hydraulic machinery runner, Proceedings of the ASME/JSME Joint Fluid Engineering Conference, ASME FED-vol. 245, FEDSM98-4867.

Peng, G., Fujikwa, S., and Cao, S. (1998b). An advanced quasi-three-dimensional inverse computation model for axial flow pump impeller design. In: *Proceedings of the XIX IAHR Symposium in Hydraulic Machinery and Cavitation*, 722–733. Singapore: World Scientific.

Peng, G., Cao, S., Ishizuka, M., and Hayama, S. (2002). Design optimization of axial flow hydraulic turbine runner part I: an improved Q3D inverse method. *International Journal for Numerical Methods in Fluids* 39 (6): 533–548.

Perie, F., and Buell, J. C. (2000). Combined CFD/CAA method for centrifugal fan simulation, The 29th International Congress and Exhibition on Noise Control Engineering, *27–30 August, Nice, France*.

Petit, O., Mulu, B., Nilsson, H., and Cervantes, M. J. (2010). Comparison of numerical and experimental results of the flow in the U9 Kaplan turbine model, Proceedings of the 25th IAHR Symposium on Hydraulic Machinery and Systems, *Timisoara, Romania*.

Pfleiderer, C. (1952). *Turbomachines*. New York: Springer-Verlag.

Powell, M.J. (1978). A fast algorithm for nonlinearly constrained optimization calculations. In: *Numerical Analysis*, 144–157. Berlin, Heidelberg: Springer.

Robert, C.P. and Casella, G. (2005). *Monte Carlo Statistical Methods*. New York: Springer.

Rodgers, C. (1980) Efficiency of centrifugal compressor impellers, AGARD Conference Proceedings, Centrifugal Compressors, Flow Phenomena and Performance, *Brussels*.

Ross, P.J. (1996). *Taguchi Techniques for Quality Engineering*. New York: McGraw-Hill.

Sale, D., Jonkman, J., and Musial, W. (2009). *Hydrodynamic Optimization Method and Design Code for Stall-Regulated Hydrokinetic Turbine Rotors*. National Renewable Energy Laboratory.

Saltelli, A., Ratto, M., and Andres, T. (2008). *Global Sensitivity Analysis: The Primer*. Hoboken, NJ: Wiley.

Samad, A. and Kim, K.Y. (2009). Surrogate based optimization techniques for aerodynamic design of turbomachinery. *International Journal of Fluid Machinery and Systems* 2 (2): 179–188.

Samad, A., Kim, K.Y., Goel, T. et al. (2008a). Multiple surrogate modeling for axial compressor blade shape optimization. *AIAA Journal of Propulsion and Power* 24 (2): 302–310.

Samad, A., Lee, K.S., and Kim, K.Y. (2008b). Multi-objective shape optimization of an axial fan blade. *International Journal of Air-Conditioning and Refrigeration* 16 (1): 1–8.

Sarraf, C., Nouri, H., Ravelet, F., and Bakir, F. (2011). Experimental study of blade thickness effects on the overall and local performances of a controlled vortex designed axial-flow fan. *Experimental Thermal and Fluid Science* 35: 684–693.

Schobeiri, M. (2005). *Turbomachinery Flow Physics and Dynamic Performance*. Berlin/Heidelberg: Springer.

Seo, S.J., Kim, K.Y., and Kang, S.H. (2003). Calculations of three-dimensional viscous flow in a multi-blade centrifugal fan by modeling blade forces. *Proceedings of Institution Mechanical Engineers Part A: Journal of Power and Energy* 217: 287–297.

Seo, S.J., Choi, S.M., and Kim, K.Y. (2008). Design optimization of a low-speed fan blade with sweep and lean. *Proceedings of Institution Mechanical Engineers Part A: Journal of Power and Energy* 222: 87–92.

Shi, J., Li, X., and Wang, S. (2014). Dynamic pressure gradient model of axial piston pump and parameters optimization. *Mathematical Problems in Engineering* Article ID: 352981 10.

Sieverding, F., Ribi, B., and Casey, M. (2004). Design of industrial axial compressor blade sections for optimal range and performance. *Journal of Turbomachinery* 126 (2): 323–332.

Singh, O.P., Khilwani, R., Sreenivasulu, T., and Kannan, M. (2011). Parametric study of centrifugal fan performance: experiments and numerical simulation. *International Journal of Advances in Engineering & Technology* 1 (2): 33–50.

Singhal, A. K., Vaidya, N., and Leonard, A. D. (1997). Multi-dimensional simulation of cavitation flows using a PDF model for phase change, Proceedings of ASME FEDSM, June 22-26, Vancouver, Canada, FEDSM97-3272.

Skotak, A., Mikulasek, J., and Obrovsky, J. (2009). Development of the new high specific speed fixed blade turbine runner. *International Journal of Fluid Machinery and Systems* 2 (4): 392–399.

Sorensen, D.N. (2001). Minimizing the trailing edge noise from rotor-only axial fans using design optimization. *Journal of Sound and Vibration* 247 (2): 305–323.

Sorensen, D.N. and Sorensen, J.N. (2000). Toward improved rotor-only axial fans – part I: a numerically efficient aerodynamic model for arbitrary vortex flow. *ASME Journal of Fluids Engineering* 122 (2): 318–323.

Sorensen, D.N., Thompson, M.C., and Sorensen, J.N. (2000). Toward improved rotor-only axial fans – part II: design optimization for maximum efficiency. *ASME Journal of Fluids Engineering* 122 (2): 324–329.

Sparlat, P. R., and Allmaras, S. R. (1994). A One-Equation Turbulence Model for Aerodynamic Flows. AIAA Paper 1992-0439.

Stauter, R.C. (1993). Measurement of the three-dimensional tip region flow field in an axial compressor. *ASME Journal of Turbomachinery* 115: 468–476.

Stein, P., Sick, M., Doerfler, P., White, P., and Braune, A. (2006). Numerical simulation of the cavitating draft tube vortex in a Francis turbine. *Proceedings of the XXIII IAHR Symposium, Yokohama, Japan.*

Stepanoff, A.J. (1948). *Centrifugal and Axial Flow Pumps: Theory, Design and Applications.* New York: Wiley.

Sturmayr, A. and Hirsch, C. (1999). Throughflow model for design and analysis integrated in a three-dimensional Navier–Stokes solver. *Proceedings of the Institution of Mechanical Engineers, Part A: Journal of Power and Energy* 213 (4): 263–273.

Sugimura, K., Jeong, S., Obayashi, S., and Kimura, T. (2008). Multi-objective robust design optimization and knowledge mining of a centrifugal fan that takes dimensional uncertainty into account, Proceedings of ASME Turbo Expo 2008: Power for Land, Sea and Air, *June 9–13, 2008, Berlin, Germany,* GT2008-51301.

Sugimura, K., Jeong, S., Obayashi, S., and Kimura, T. (2009). Kriging-model-based multi-objective robust optimization and trade-off rule mining of a centrifugal fan with dimensional uncertainty. *Journal of Computational Science and Technology* 3 (1): 196–211.

Sugimura, K., Obayashi, S., and Jeong, S. (2010). Multi-objective optimization and design rule mining for an aerodynamically efficient and stable centrifugal impeller with a vaned diffuser. *Engineering Optimization* 42 (3): 271–293.

Sun, J. and Elder, R.L. (1998). Numerical optimization of a stator vane setting in multistage axial-flow compressors. *Proceedings of the Institution of Mechanical Engineers, Part A: Journal of Power and Energy* 212 (4): 247–259.

Susan-Resiga, R., Ciocan, G.D., Anton, I., and Avellan, F. (2006). Analysis of the swirling flow downstream a Francis turbine runner. *ASME Journal of Fluids Engineering* 128: 177–189.

Susan-Resiga, R., Muntean, S., Anton, I., and Bernad, S. (2003). Numerical investigation of 3D cavitating flow in Francis turbines, Conference on Modelling Fluid Flow (CMFF'03) The 12th International Conference on Fluid Flow Technologies, *Yokohama, Japan.*

Thakur, S., Lin, W., and Wright, J. (2002). Prediction of flow in centrifugal blower using quasi-steady rotor–stator models. *Journal of Engineering Mechanics* 128 (10): 1039–1049.

Tiwari, S., Koch, P., and Fadel, G. (2008). AMGA: an archive-based micro genetic algorithm for multi-objective optimization. In: *GECCO Conference,* 729–736. Atlanta, GA, July 12–16.

Tong, S.S. and Gregory, B.A. (1990, June). Turbine preliminary design using artificial intelligence and numerical optimization techniques. In: *ASME 1990 International Gas Turbine and Aeroengine Congress and Exposition.* New York: American Society of Mechanical Engineers.

Tridon, S., Ciocan, G. D., Barre, S., and Tomas, L. (2008). 3D time-resolved PIV measurement in a francis turbine draft tube, Proceedings of the 24th Symposium on Hydraulic Machinery and Systems.

Tridon, S., Barre, S., Ciocan, G.D., and Tomas, L. (2010). Experimental analysis of the swirling flow in a Francis turbine draft tube: focus on radial velocity component determination. *European Journal of Mechanics B/Fluids* 29 (4): 321–335.

Trivedi, C., Cervantes, M.J., Gandhi, B.K., and Dahlhaug, O.G. (2013). Experimental and numerical studies for a high head Francis turbine at several operating points. *ASME Journal of Fluids Engineering* 135: 111102.

Tsuei, H.H., Oliphant, K., and Japikse, D. (1999). *The Validation of Rapid CFD Modeling for Turbomachinery*. London: Institution of Mechanical Engineers.

Tsurusaki, H., Imaichi, K., and Miyake, R. (1987). A study on the rotating stall in vaneless diffusers of centrifugal fans. *JSME International Journal* 30 (260): 279–287.

Vacca, A. and Cerutti, M. (2007). Analysis and optimization of a two-way valve using response surface methodology. *International Journal of Fluid Power* 8 (3): 43–59.

Vande Voorde, J., Dick, E., Vierendeels, J., and Serbmyns, S. (2004). Performance prediction of centrifugal pumps with steady and unsteady CFD-methods. In: *Advances in Fluid Mechanics IV* (ed. M. Rahman, R. Verhoeven and C.A. Brebbia), 559–568. Southampton, UK: WIT Press.

Velarde-Suarez, S., Ballesteros-Tajadura, R., Santolaria-Morros, C., and Gonzalez-Perez, J. (2001). Unsteady flow pattern characteristics downstream of a forward-curved blades centrifugal fan. *ASME Journal of Fluids Engineering* 123: 265–270.

Veres, J.P. (1994). Centrifugal and axial pump design and off-design performance prediction. *NASA Techincal Memorandum* 106745: 1–24.

Von Backstrom, T. W., and Roos, T. H. (1993). The streamline throughflow method for axial turbomachinery flow analysis. Presented at the Eleventh International Symposium on Air Breathing Engines, Tokyo, Japan, pp. 347–354.

Wallis, R.A. (1961). Axial Flow Fans. In: *Design and Practice*. London: George Newnes Limited.

Wang, W., Zhang, L., Yan, Y., and Guo, Y. (2007). Large-eddy simulation of turbulent flow considering inflow wakes in a Francis turbine blade passage. *Journal of Hydrodynamics, Series B* 19 (2): 201–209.

Wang, L.Q., Yin, J.L., Jiao, L. et al. (2011). Numerical investigation on the "S" characteristics of a reduced pump turbine model. *Science China Technological Sciences* 54 (5): 1259–1266.

Wei, N. (2000). Significance of Loss Models in Aerothermodynamic Simulation for Axial Turbines. PhD Thesis, Department of Energy Technology, Division of Heat and Power Technology, Royal Institute of Technology, Sweden.

Whitfield, A. and Baines, N.C. (2002). *Design of Radial Turbomachines*. Harlow, Essex: Longman.

Williams, J.E.F. and Hawkings, D.L. (1969). Sound generation by turbulence and surfaces in arbitrary motion. *Philosophical Transactions of the Royal Society of London Series A* 264 (1151): 321–342.

Wisler, D.C. and Mossey, P.W. (1973). Gas velocity measurements within a compressor rotor passage using the laser Doppler velocimeter. *ASME Journal of Engineering for Power* 95 (2): 91–97.

Witten, I.H. and Frank, E. (2005). *Data mining*, 189–199. San Francisco: Morgan Kaufmann, ch. 6.

Wright, T. and Simmons, W.E. (1990). Blade sweep for low-speed axial fans. *Journal of Turbomachinery* 112 (1): 151–158.

Wu, C.H. (1952). *A general theory of three-dimensional flow in subsonic and supersonic turbomachines of axial-, radial, and mixed-flow types (No. NACA-TN-2604)*. Washington DC: National Aeronautics and Space Administration.

Wu, J., Shimmei, K., Tani, K. et al. (2007). CFD based design optimization for hydro turbines. *ASME Journal of Fluids Engineering* 129: 159–168.

Wu, Y., Liu, J., Sun, Y. et al. (2013). Numerical analysis of flow in a Francis turbine on an equal critical cavitation coefficient line. *Journal of Mechanical Science and Technology* 27 (6): 1635–1641.

Xiao, Y.X., Sun, D.G., Wang, Z.W. et al. (2012). Numerical analysis of unsteady flow behavior and pressure pulsation in pump turbine with misaligned guide vanes. *IOP Conference Series: Earth and Environmental Science* 15 (3): 032043–032051.

Xiao, Y., Wang, Z., Zhang, J., and Luo, Y. (2014). Numerical predictions of pressure pulses in a Francis pump turbine with misaligned guide vanes. *Journal of Hydrodynamics* 26 (2): 250–256.

Yamazaki, S. (1986). An experimental study on the aerodynamic performance of multi-blade blowers (1st report). *Transactions of JSME(B)* 52 (484): 3987–3992.

Yamazaki, S. (1987a). An experimental study on the aerodynamic performance of multi-blade blowers (2nd report). *Transactions of JSME(B)* 53 (485): 108–113.

Yamazaki, S. (1987b). An experimental study on the aerodynamic performance of multi-blade blowers (3rd report). *Transactions of JSME(B)* 53 (490): 1730–1735.

Yang, W. and Xiao, R.F. (2014). Multiobjective optimization design of a pump–turbine impeller based on an inverse design using a combination optimization strategy. *ASME Journal of Fluids Engineering* 136: 014501.

Yang, L., Hua, O., and Zhao-Hui, D. (2007a). Optimization design and experimental study of low-pressure axial fan with forward-skewed blades. *International Journal of Rotating Machinery* 2007, Article ID: 85275: 10. https://doi.org/10.1155/2007/85275.

Yang, L., Ouyang, H., and Zhao-Hui, D.U. (2007b). Experimental research on aerodynamic performance and exit flow field of low pressure axial flow fan with circumferential skewed blades. *Journal of Hydrodynamics* 19 (5): 579–586.

Yang, L., Jie, L., Hua, O., and Zhao-Hui, D. (2008). Internal flow mechanism and experimental research of low pressure axial fan with forward-skewed blades. *Journal of Hydrodynamics* 20 (3): 299–305.

Yin, J.L., Liu, J.T., Wang, L.Q. et al. (2010). Performance prediction and flow analysis in the vaned distributor of a pump turbine under low flow rate in pump mode. *Science China Technological Sciences* 53 (12): 3302–3309.

Yin, J., Wang, D., Wei, X., and Wang, L. (2013). Hydraulic improvement to eliminate S-shaped curve in pump turbine. *ASME Journal of Fluids Engineering* 135: 0711105.

Yiu, K.F.C. and Zangeneh, M. (1998, June). A 3D automatic optimization strategy for design of centrifugal compressor impeller blades. In: *ASME 1998 International Gas Turbine and Aeroengine Congress and Exhibition*. American Society of Mechanical Engineers.

Younsi, M., Bakir, F., Kouidri, S., and Rey, R. (2007). Influence of impeller geometry on the unsteady flow in a centrifugal fan: numerical and experimental analyses. *International Journal of Rotating Machinery* 2007, Article ID: 34901: 10.

Yu, Z., Li, S., He, W. et al. (2005). Numerical simulation of flow field for a whole centrifugal fan and analysis of the effects of blade inlet angle and impeller gap. *HVAC & R Research* 11 (2): 263–283.

Yu, Q., Cai, M., Shi, Y., and Fan, Z. (2014). Optimization of the energy efficiency of a piston compressed air engine. *Journal of Mechanical Engineering* 60 (6): 395–406.

Zangeneh, M., Goto, A., and Harada, H. (1999). On the role of three-dimensional inverse design methods in turbomachinery shape optimization. *Proceedings of the Institution of Mechanical Engineers, Part C: Journal of Mechanical Engineering* 213 (1): 27–42. https://doi.org/10.1243/0954406991522167.

Zangeneh, M., Vogt, D., and Roduner, C. (2002). Improving a Vaned Diffuser for a Given Centrifugal Impeller by 3D Inverse Design, ASME Paper GT-2002-30621.

Zhang, R., Mao, F., Wu, J.Z. et al. (2009). Characteristics and control of the draft-tube flow in part-load Francis turbine. *ASME Journal of Fluids Engineering* 131: 021101.

Zhou, D., Zhou, J., and Song, J. (1996). Optimization Design of an axial-flow fan used for mining local-ventilation. *Computers & Industrial Engineering* 31 (3/4): 691–696.

Zhu, X., Lin, W., and Du, Z. (2005). Experimental and numerical investigation of the flow field in the tip region of an axial ventilation fan. *ASME Journal of Fluids Engineering*.

Zobeiri, A., Kukny, J-L., Farhat, M., and Avellan, F. (2006). Pump-turbine rotor-stator interactions in generating mode: pressure fluctuation in distributer channel, 23rd IAHR Symposium, *Yokohama, Japan*.

5

Optimization of Fluid Machinery for Renewable Energy Systems

Renewable sources of energy are those that are replenished naturally on usage. Wind, tides, waves, sunlight, ocean currents, geothermal heat, biomass, salinity gradient and so on constitute the major renewable energy sources. Renewable energy sources are generally clean, emissions-free power production resources. A few renewable energy resources are discussed in this chapter. Among them, many use turbomachinery in one form or other. For example, for night use of solar power, a pump is used to store energy in a higher location and a turbine extracts energy at night. The pump or the turbine is basically a turbomachine.

5.1 Wind Energy

Wind turbines are a very special class of turbomachines. Their use and related engineering topics, including aerodynamic design and optimization, has gained increasing worldwide interest over the last decade due to the exponential growth of wind energy exploitation. In fact, among all the so-called renewables, wind turbine technology has seen the fastest growing rate in terms of global installed power (from 17 GW in 2000 to more than 430 GW in 2015 [Council 2016], no less than 30.8 GW just in China).

Wind turbine technology is dominated by very large open rotors of the horizontal-axis type (Horizontal-Axis Wind Turbines – HAWTs). In these machines, the wind flows across a rotor, usually featuring three blades, whose angular velocity is in the same direction of the blowing wind (Spera 1994). It is estimated that more than 95% of the energy produced from wind worldwide comes from HAWTs. The remaining part is dealt with using vertical axis wind turbines (VAWTs), usually slightly less efficient and more expensive compared to HAWTs, but still under investigation for particular applications including distributed generation, building integration, and small power production in urban environments. In VAWTs, angular velocity is somewhat perpendicular to the wind direction and this has proven to give some potential benefits over HAWTs in turbulent winds and in sites characterized by rapid variations in wind directions. Despite this, VAWT technology has received little interest so far in practical applications, while it is being significantly studied at research and academic level. Regardless the type, the most important quantities with which every wind turbine is judged are almost invariably (i) the (annual) cost per unit of energy produced (COE) and (ii) the amount of (annual)

Design Optimization of Fluid Machinery: Applying Computational Fluid Dynamics and Numerical Optimization, First Edition. Kwang-Yong Kim, Abdus Samad, and Ernesto Benini.
© 2019 John Wiley & Sons Singapore Pte. Ltd. Published 2019 by John Wiley & Sons Singapore Pte. Ltd.

energy produced (AEP). Both are strongly dependent on the geometry of the turbine and, in turn, on turbine aerodynamic efficiency. It is common practice to evaluate such efficiency by comparing the turbine actual power produced to the maximum theoretical power within the stream tube of air intercepting the turbine rotor that, in fact, is the kinetic power of the incoming wind:

$$C_p = \frac{P}{0.5\rho A U^3} \tag{5.1}$$

where P = power captured, ρ = air density, A = rotor swept area, and U = ambient air speed.

The aforementioned quantity is called the turbine power coefficient. Actually, there are certain limits on the maximum attainable values of the power coefficient and these are dependent on the type of wind turbine. In a conventional HAWT, simple momentum exchange balances across the swept area indicate that the maximum theoretical C_p is slightly higher than 59%. Actual peak values for C_p are slightly higher than 0.50 for modern multi-MW turbines (today, the highest claimed C_p is 0.52 while staying at 0.48–0.50 on average for very large machines) and in the range of 0.35–0.40 for most VAWTs, that is, at least 20% lower.

Both COE and AEP are functions of C_p, which in turn is a non-trivial function of the wind speed. Rather than the wind speed, it is most meaningful to account for C_p variations as a function of the so-called tip-speed ratio, that is, the ratio between the rotor tip velocity and the wind speed:

$$\lambda = \frac{U_{tip}}{U} \tag{5.2}$$

As the C_p varies with the wind speed and the latter changes over time by definition, the annual energy production in a specific site is used as an effective measure of turbine efficiency. In order to calculate the AEP, a wind speed distribution (usually of the Weibull type) must be provided that is particular to the site in which the turbine will be installed. A continuous probability density Weibull function is defined as:

$$f_W = \left(\frac{k}{c}\right)\left(\frac{U}{c}\right)^{k-1} \exp\left[-\left(\frac{U}{c}\right)^k\right] \tag{5.3}$$

where c is the scale parameter in m s^{-1}, and k is the Weibull curve dimensionless shape parameter. For a given site, the Weibull function can be calculated over a discrete period of time Δt_i. As a result, the AEP is calculated using the following formula, where n stands for the number of time steps to over one year of time:

$$AEP = \sum_{i=1}^{n} P_i f_W(U_i) N \Delta t_i = \sum_{i=1}^{n} C_{p,i} 0.5 \rho A U_i^3 f_W(U_i) N \Delta t_i \tag{5.4}$$

To achieve more general design indications, the influence of rotor radius and rated power should be investigated during the optimization process. When this is the case, maximization of AEP could indeed drive the search toward very large rotors with very thin blades, thus making the results of limited interest in engineering because of their infeasibility. Instead of using AEP alone, AEP per unit of the rotor swept area (proportional to AEP/R^2, R being the rotor tip radius), should be employed in order to obtain more general and effective design guidelines.

An accurate estimation of the Cost of Energy is rather difficult to achieve. A simple yet effective cost model is calculated using the following expression given by Giguère et al.:

$$COE = \frac{TC + BOS}{AEP} FCR + O\&M \, [\$/kWh] \tag{5.5}$$

where *TC* is the turbine cost proportional to blade weight, *BOS* is the balance of station proportional to turbine rated power, *FCR* is the fixed charge rate, and *O&M* is the cost of operation and maintenance. Several sources indicate the current values for those figures. The cost model by Giguère is based on the assumption that total turbine cost can be reconstructed on the basis of the cost of turbine blade alone. This requires that, for a given rated power, the proportion among turbine component costs does not change. However, it does not consider the possible variations of such proportions that might occur for very different turbine designs, for example, in the case of very different tip speeds. In fact, the tip speed influences the maximum torque needed to achieve a prescribed rated power, which in turn affects the design and cost of the drive train. In this case, the cost model does not capture such effects. From the available information to the public, Malcolm (2003) calculated the rotor mass per unit of swept area for some commercial HAWTs and VAWTs. He reported that, for a given swept area, VAWTs are almost 10 times higher in mass compared to HAWTs, regardless of the turbine size. As Malcolm explains, however, these figures do not take the tower weight into account for HAWTs, but still a factor of about four exists between vertical and horizontal mass/swept area.

On the other hand, unitary masses tend to increase as a square-cube function of the swept area of both turbine types (scale-up law), which means that the energy costs related to turbine mass should favor HAWTs and, among these, the ones of small power produced. However, as pointed out by Malcolm, installation costs are much higher for single small wind turbines, and the machine cost is a greater fraction of the final installed cost than for large machines in wind farms. This argument has been in fact confirmed by other researchers who found that COE increases with turbine-rated power. Therefore, the recently developed large HAWTs would be less suitable for two-dimensional layout of a wind farm than for stand-alone installations, in which AEP represents the real energetic objective.

In offshore applications, the potential use of very large VAWTs seems to offer a unique advantage over HAWTs. In fact, as scaling up does affect turbine mass/swept area in almost the same way between the two types, VAWTs should take advantage of their better behavior against gravity loads. These loads seem to limit the design of very large HAWT rotors. However, the concrete feasibility of very large offshore VAWTs still requires further studies before it can be validated. The DeepWind Project funded by the European Union was an example of proving this concept using tilting rotors (The DeepWind Project 2016).

5.1.1 Optimization of Horizontal-Axis Wind Turbines

The aerodynamic design of a HAWT rotor is a complex procedure characterized by several trade-off decisions aimed at finding the optimum overall performance and economy. The decision-making process is difficult and the design trends are not uniquely established: a number of different commercial turbine types in use today have been derived from both theoretical and empirical methods, but there is no clear evidence on which of these have to be regarded as optimal. The reason for this is that HAWTs have to be optimized for a specific wind site.

For a given Weibull wind distribution, the designer might want to know which rotor gives the maximum AEP density for a given (or target) COE. On the other hand, the designer would benefit from achieving the minimum COE for a desired AEP density. In general, the knowledge of how AEP varies as a function of COE for a given set of HAWTs is fundamental for exploiting a windy site. Using the well-known terminology of optimization, this is equivalent to search for the set of Pareto-optimal design solutions with respect to AEP and COE. Other non-aerodynamic objective functions could be also of extreme interest in optimizing HAWT, such as maximizing blade rigidity and/or optimize turbine dynamics.

To achieve these tasks, first of all some flow models must be constructed. These include a variety of methods, from Blade Element Methods (BEMs) to computational fluid dynamics (CFD). BEMs are still today preferred to CFD in optimization problems thanks to their inexpensive use and easiness in coupling with other design disciplines, such as structural dynamics. Despite this, BEMs provide accurate performance predictions if proper corrections are implemented.

5.1.2 Blade Element Methods

These are often referred to as "strip theories" and they combine two basic theories: (i) the momentum balance on annular stream tubes passing through a turbine rotor (giving rise to the elementary components of axial and tangential thrust) and (ii) force balance generated by the airfoil lift and drag at various sections along the blade. By equating the force components arising from the two approaches at each radial station (or blade strip), a system of nonlinear equations can be obtained and solved iteratively (Hansen 1993). The rotor is divided into a finite number of annular control volumes (the "strips," see Figure 5.1), each independent from the others, where the force exerted by the air on the blades is considered constant in each annular element.

Despite of its simplicity and its oldness (strip theory is an extension of the actuator disk theory first proposed by Rankine and Froude in the late nineteenth century), BEM codes

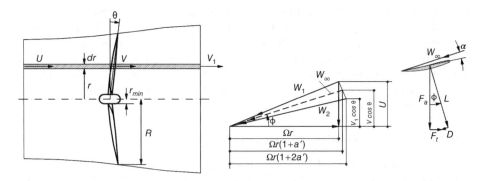

Figure 5.1 Control volume used in BEM models (left) and velocity triangles at an arbitrary radius r of a Horizontal-Axis Wind Turbine (right). Source: Reproduced with permission from Benini and Toffolo 2002 (figures 1 and 2 from original source). Copyright © 2002 by American Society of Mechanical Engineers.

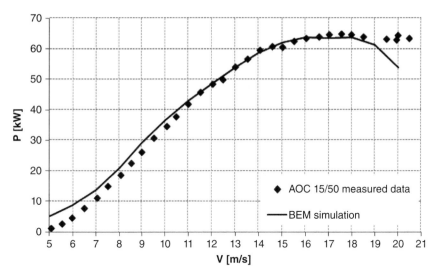

Figure 5.2 Comparison between numerical and experimental power curves for the AOC 15/50 wind turbine. Source: Reprinted from Dal Monte et al. 2017 (figure 3 from original source) with permission from Elsevier.

prove to be very reliable provided that some corrections are implemented, such as tip and hub-loss models, Glauert correction to account for large induced velocities, accurate extended lift and drag polars, rotational, and cascade effects. Other, more sophisticated corrections, such as those related to three-dimensional inboard stall delay effects have been proposed and can be used if necessary (Sant 2007, Lindenburg 2003). The principal advantage of BEM codes compared to more advanced tools (like CFD), relies in its computational inexpensiveness and ease of implementation. When corrections are adequately modeled and airfoil polars accurately predicted, BEM codes can be very helpful for both design and optimization purposes. It is also worth recalling that, since experimental polars are often not available in the design phase, the utilization of numerical codes based on corrected panel methods, such as XFoil (Lindenburg 2003) or RFoil (Drela 1989), can be of priceless help in the prediction of actual airfoil characteristics.

As an example, fig displays a comparison between the experimental and predicted power versus air speed curve for the AOC 15/50 wind turbine installed at NREL's National Wind Technology Center (NWTC) in Colorado (Jacobson et al. 2003). Numerical predictions have been obtained by the authors using a BEM code where polar curves for the turbine airfoils were obtained using an RFoil and extended up to 90° of incidence angle using the Lindenburg method. As apparent from the Figure 5.2, a remarkable agreement can be observed up to a wind velocity of 18 m s^{-1}, where a deep blade stall occurs. From this wind speed on, the results of the BEM code cannot be regarded as accurate; however, the turbine design and optimization is usually carried out in operating conditions fairly away from the rotor stall.

5.1.3 Turbine Parameterization

In optimization problems involving wind turbines, parameterization usually refers to rotor geometrical shape, although functional parameters can be considered as well.

Using a BEM code, it is common practice to use either a part or the entire set of the following decision variables (Benini and Toffolo 2002):

- Tip speed ΩR. A range $40 < \Omega R < 80\text{ m s}^{-1}$ is usual, according to the values usually adopted in the practice. Both Ω and R are then obtained by imposing a fixed turbine rated power P. Today's turbines typically run at an angular velocity such that the tip speed stays quite below 80 m s^{-1} in order to control the maximum allowable stress at blade root as well as the noise emission.
- Number of blades. In HAWTs for power production, blade number is between one and three.
- Hub/tip ratio $v = r_{min}/R$. There is a lack of knowledge in the open literature regarding this variable, although it is widely recognized that it has a pre-eminent role for determining the operating range of the turbine, since it is linked to the appearance of the stall phenomenon in the airfoils at the root. A range $0.05 < v < 0.2$ is typically assumed.
- Chord distribution along the blade $c/R = f(r/R)$. The chord length is defined as a function of blade radius. Bézier curves can be used to this purpose. Four control points is considered a good compromise between the necessity for blade shape flexibility and the constraints dictated by blade manufacturability (Figure 5.3).
- Twist distribution along the blade $\gamma_c = f(r/R)$. The twist is defined as a function of blade radius in the same way as the chord distribution (Figure 5.3). Twist distribution can be rather arbitrary although it has to conform to the correct incidence of the relative velocity along the blade radius.
- Shell (skin) thickness distribution along the blade. The distribution can be parameterized using Bézier or other types of curves (in Figure 5.4 a parabolic distribution has been used), the maximum thickness being located at blade root. The actual value of hub thickness is determined by structural considerations, in particular considering bending stress due to blade loading and tensile stress due to centrifugal forces in the hub profile. A very simple model can be used to determine the maximum equivalent normal stress as follows:

$$\sigma_{max} = \chi \left(\frac{M_b}{I(s)} \frac{0.21 c_{hub}}{2} + \frac{F_c}{S} \right) \tag{5.6}$$

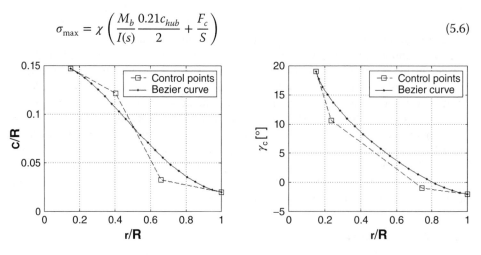

Figure 5.3 Bézier curves describing chord and twist distributions. Source: Reproduced with permission from Benini and Toffolo 2002 (figure 4 from original source) Copyright © 2002 by American Society of Mechanical Engineers.

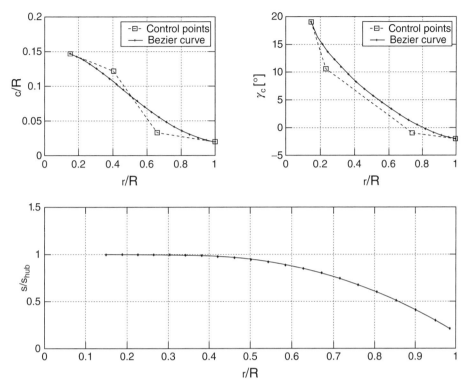

Figure 5.4 Parabolic curve describing shell thickness distributions. Source: Reproduced with permission from Benini and Toffolo 2002 (figure 5 from original source) Copyright © 2002 by American Society of Mechanical Engineers.

where χ is a safety margin, M_b is the bending moment, I is the moment of inertia of the hub profile, F_c is the centrifugal force, and S is the contour area of the hub profile.

- Coning angle θ. It is used for structural reasons and for the accomplishment of an acceptable accommodation for the tower rather than for aerodynamic purposes. A specific value of θ is usually assumed.
- Tilt angle δ, that is, the angle between wind direction and rotor axis of rotation. It is often used for tower accommodation purposes. Also in this case, a specific value for δ is assumed.
- Airfoil properties at any radius (C_L and C_D as functions of the incidence angle δ). The airfoil family is made up of three airfoils (root, primary, and tip airfoils) and is normally considered to realize good aerodynamic performance and conform to shape constraints. For instance, as most wind turbine blades have a circular section that attaches to the hub, a smooth shape transition from the circular section to the root airfoil section is needed and this requires the root airfoils to be relatively thick relative to the chord length. Because thick airfoils do not show great efficiency, a primary thinner airfoil is then accommodated, which is stacked over the root to 75% of the blade span assuring a greater lift over drag ratio. A tip airfoil, usually much thinner than the primary one, is finally positioned to 95% of the blade span. Above 95%, tapering occurs in order to minimize aerodynamic tip losses.

The shape distribution of these airfoils along the blade and their thickness can be parameterized. Early studies on wind turbine optimization considered fixed blade shapes and concentrated on blade and twist distribution as decision variables (Selig and Coverstone-Carrol 1996). More recent studies included airfoil shapes as part of the optimization process. In Figure 5.5, an example of airfoil parameterization using Bézier curves is given along with the pertinent polar curves obtainable using the RFoil code. In order to have a good representation of typical wind turbine geometry, a total of 10 control points is advisable: five for the suction side plus five for the pressure side corresponding to the x coordinates in Figure 5.5. Note that in this case the position of the leading edge is fixed while the actual trailing edge location must obey to the spanwise chord distribution. If three airfoil families are used, a total of $10 \times 3 = 30$ decision variables for describing airfoils' shapes are suggested.

5.1.4 Strategies for Rotor Optimization

As explained in (Dal Monte 2017), an optimization of either AEP/R^2 or COE, or both, involving all the previously mentioned decision variables at the same time usually provides suboptimal results due to excessive complication of the fitness landscape. More significant engineering results can be obtained using a sequential approach as described next.

1. Chord and twist distribution are optimized first while maintaining the fixed airfoil shape.
2. Hub profile shape is then optimized, keeping the results of step 1.
3. Primary profile shape is optimized next, keeping the results of step 2.
4. Tip profile shape is finally optimized while keeping the results of all the previous steps.

Using such an approach, improvements in the AEP/R^2 in the order of 10% compared to a standard wind turbine configuration can be sought for (Dal Monte 2017). Among the interventions here, optimization of chord and twist (step 1) along with optimization of tip profile shape (step 4) provide the majority of the increment in the total energy production. Regarding COE, the expected improvements are in the order of a 2–4% reduction, a less evident result compared to the obtainable maximizing AEP, since most commercial turbines are very close to the minimum COE, and also because the cost model of Eq. (5.9) is fairly simple and might not capture sensible variations in the turbine cost other than those related to the mass of the blade.

5.2 Ocean Energy

The ocean is considered a vast source of energy. Ocean energy consists of energy that can be retrieved from ocean waves, tides, currents, salinity, and thermal gradients. Ocean energy and its extraction technologies are explained in detail in the following sections.

Oceans cover up to 70% of the Earth's surface area. Oceans possess a vast amount of renewable energy potential in different forms. It is predicted that the ocean's energy

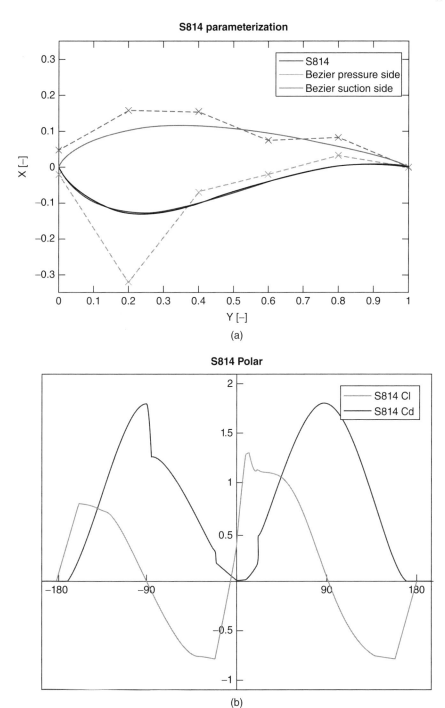

Figure 5.5 (a) Parameterization of a wind turbine airfoil and (b) extended airfoil polars (Dal Monte 2017).

potential well exceeds our present energy needs. These can be harnessed by devising various techniques and can be converted to a convenient form of electricity.

Energy is extracted from the following natural phenomena observed in oceans for various reasons like thermal radiation from the Sun, gravitational pull of the Moon, rotation of the Earth, and so on. The following paragraphs brief about the natural phenomena used to produce ocean energy.

5.2.1 Temperature Gradients

Since oceans cover a large surface area of the Earth, this makes them the world's biggest solar collectors. The surface of the ocean is more heated compared to deep ocean water. This temperature difference in the ocean water provides the potential for thermal energy. This difference in temperature between the cooler deep and warmer shallow seawaters is trapped to operate a heat engine to produce energy.

5.2.2 Tides and Tidal Currents

Tidal currents occur in conjunction with the high and low tides caused by the gravitational forces of the Moon, the Sun, and also the Earth's rotation. When a tide occurs, the vertical movement of the surface water near the shore leads the water to move in the horizontal direction, creating currents. These tidal currents can occur in two opposite directions. When water moves toward the land, it "floods" or "flows", and when it moves away from the land, it "ebbs." During high tide, the flood current has greater intensity than the ebb current and during low tide, the intensity of the ebb current becomes more than that of the flood current.

5.2.3 Salinity Gradients

Energy can be retrieved by employing suitable methods from the variance in salt concentration between river water and seawater. From these naturally occurring phenomena, energy can be harnessed into a useful form by devising suitable technologies.

5.2.4 Waves

Waves are produced due to the winds moving over the ocean surface. The cause of winds is the differential heating of the air above the ocean surface. Wave energy can be captured and effectively used for desalination of water or power production. The energy thus harnessed from the oceans is known as the ocean energy. The technologies devised to harness this energy are termed ocean energy technologies.

5.3 Energy Extraction from Ocean Waves

The equipment used for wave energy conversion is known as a wave energy converter. A large variety of wave energy converters are currently available. They can be classified

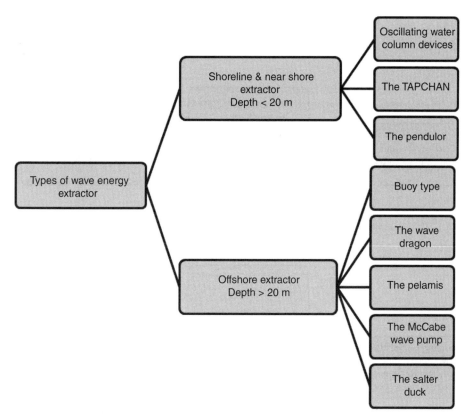

Figure 5.6 Classification of WEC devices characterized based on the depth of installation. Source: Reproduced with permission from Shehata et al. 2017 (figure 1 from original source) © Wiley-VCH Verlag GmbH & Co. KGaA.

according to the depth at which they are installed at sea (deep, intermediate, shallow), the distance from the shore (offshore, shoreline, near-shore), and the wave interaction with respective motions (heaving, surging, pitching) (Lewis et al. 2011).

Figure 5.6 shows the different classifications of Wave Energy Converter (WEC) devices characterized on depth and distance from shore. Figure 5.7 elucidates the classification of wave energy converting devices on the basis of working methodology.

5.4 Oscillating Water Column (OWC)

The OWC is the most commonly used wave energy equipment. An OWC consists of a chamber given an opening to the sea underneath the waterline. Water enters the compartment and air present inside the chamber is pressurized within the chamber as waves approach the OWC device. The pressurized air is then discharged to the atmosphere through a turbine. As water evacuates the chamber, air is drawn into the chamber through the turbine (Figure 5.8). Equipment can be classified into fixed structure OWC and floating-structure OWC.

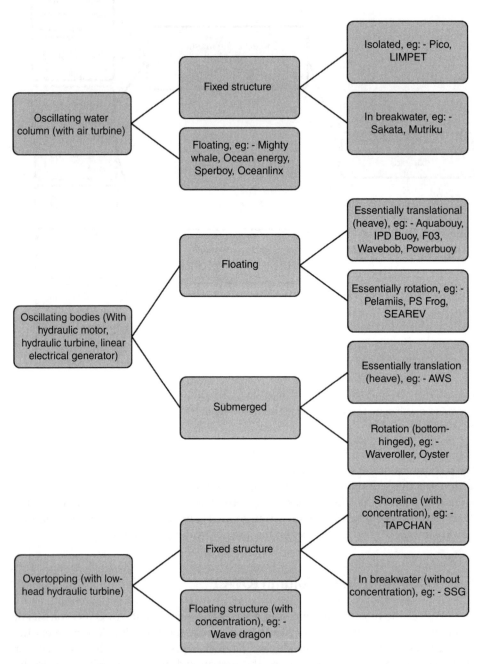

Figure 5.7 Classification of WEC devices characterized based on working principles. Source: Reprinted from Antonio 2010 (figure 4 from original source) with permission from Elsevier.

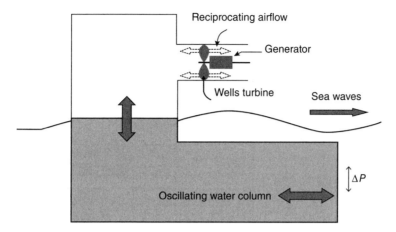

Figure 5.8 OWC operation. Source: Reprinted from Ceballos et al. 2015 (figure 1 from original source) with permission from Elsevier.

5.4.1   Fixed-Structure OWC

A fixed structure OWC stands on the bottom of the sea or is fixed to a steady cliff. They are generally placed near the shore or on the shoreline. If placed near the shore, they have a number of advantages like easy installation and maintenance, and the absence of deep-water moorings and lengthy underwater cables (Falcão and Gato 2012).

5.4.2   Floating-Structure OWC

These are not fixed to the shore or at the bottom of the sea. Instead, they are floating structures that move up and down on the sea surface. The floating OWC devices are loosely chained to the seabed and are permitted to oscillate. The oscillations of the OWC with respect to the ocean waves create pneumatic pressure differences in the air chamber arrangement, which drives an air turbine.

5.5   Classification of Turbines

Alan Arthur Wells of Queen's University, Belfast designed the Wells turbine in the late 1970s. The Wells turbine uses symmetrical aerofoils and is a bidirectional turbine. It is easier to maintain and is economical as there are no moving components except for the main turbine rotor. Conversely, the aerofoil's high angle of attack generates more drag. As a result, some of its efficiency is sacrificed at high airflow rates. At high angle of attack the turbine stalls result in low efficiency. The turbine is efficient at low-speed airflows. There are two main types of air turbine used in OWC: the impulse turbine and Wells turbine. Figure 5.9 shows the different kinds of turbines used for OWC. The impulse turbine requires guide vanes to be placed before and after the rotor to maintain the unidirectional rotation of the same. The guide vanes reduce the turbine performance.

5.5.1   Wells Turbine

The Wells turbine (Figure 5.10) is considered the simplest and most commonly used self-rectifying device for wave energy conversion. It absorbs the axial force from the

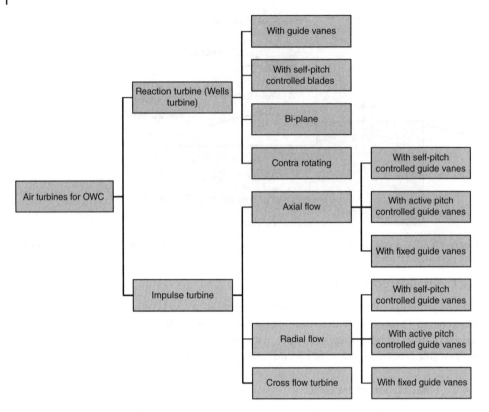

Figure 5.9 Classification of air turbines.

incoming air while the tangential force causes the turbine to rotate. For a symmetrical aerofoil, the direction of tangential force is the same for both positive and negative values of the angle of attack (α). Therefore, the direction of rotation of the rotor is always independent of the direction of the airflow.

If the absolute flow velocity is C and the tangential rotor velocity is U, then the relative flow velocity (W) is the result of C and U. According to the two-dimensional cascade theory, the undisturbed flow velocity (W) is the average of the upstream and downstream relative velocities (W_1 and W_2, respectively), and forms an angle α with respect to the blade chord (Torresi et al. 2009) (Figure 5.11). According to classical aerofoil theory, an aerofoil set at an angle of incidence α in a fluid flow generates a lift force, L, normal to the free stream. The aerofoil also experiences a drag force, D, in the direction of the free stream. The lift, L, and drag, D, forces (perpendicular and parallel to W, respectively) can be resolved into the tangential and axial components (F_u and F_n, respectively) whose magnitudes vary during the cycle (Raghunathan and Abtan 1983). Equations (2.3) and (2.4) express both the components.

$$F_u = L\sin\alpha - D\cos\alpha \tag{5.7}$$

$$F_n = L\sin\alpha + D\cos\alpha \tag{5.8}$$

Figure 5.10 Schematic of a Wells turbine. Source: With kind permission from (figure 1 from original source), Soltanmohamadi and Lakzian 2015.

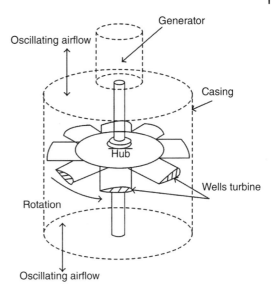

Figure 5.11 Velocity diagrams and forces acting on a blade. Source: Reproduced with permission from Torresi et al. 2009 (figure 2 from original source), Copyright © 2009 by American Society of Mechanical Engineers.

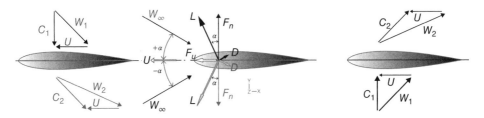

The force in the tangential direction (F_u) is identical for both negative and positive values of α in a symmetrical airfoil. When these airfoil blades are set about a rotation axis, their rotation will be in the direction of F_u regardless of the airflow direction. Thus, the direction of F_u is not dependent of the axial flow direction. Whenever the drag becomes dominant, F_u will become negative. This occurs whenever the flow rate reaches zero or when stall conditions arise.

L and D increase with the rise in α for real fluids up to a limit after which the flow around the airfoil body separates. The stall angle can be defined as the incidence angle where the flow separates from the surface of the aerofoil. As α increases beyond the stall angle, the lift decreases and drag increases significantly.

However, there are certain disadvantages inherent with the Wells turbine, such as its poor efficiency and starting characteristics, large noise levels, and comparatively narrower operating range. The operating range is bounded by the stall angle. The inability of the turbine to reach its operating speed due to its poor starting characteristics is termed crawling (Raghunathan and Tan 1982).

5.5.2 Impulse Turbine

Impulse turbines utilize the impulse force created on the blades when a high velocity stream/jet of fluid impinges on them. After the impulse, the flow direction of the stream is changed, energy is transferred to the blade and it leaves the blade with a very low kinetic energy. The velocity head of the fluid upstream has to be maximized for a high power output. Therefore, the pressure head of the fluid upstream is converted into velocity head by accelerating the fluid through nozzles. By the passage of the fluid through nozzles, the pressure head of the fluid is converted into a velocity head before reaching the blades. In an impulse turbine, the energy transfer takes place only because of the impulse and there is negligible pressure drop at the blade.

The torque produced by an impulse turbine is proportional to the change in momentum of the fluid flowing over the turbine rotor, as per the Euler turbo machinery equation (Dixon and Hall 2014). When a one-dimensional approximation is applied, the equation can be expressed as,

$$T = \dot{m}\,(r_1 V_1 - r_2 V_2) \tag{5.9}$$

If $r_1 = r_2 = r$, the equation can be written as,

$$E = \omega r\,(V_1 - V_2) \tag{5.10}$$

For a positive torque, the inlet velocity should be higher than the outlet velocity, that is, $V_1 > V_2$. The nozzles are placed in the inlet side so as to accelerate the incoming stream of fluid to a higher velocity. In an OWC, the airflow is bidirectional. The stators are positioned on both sides of the rotor such that the turbine is unidirectional regardless of the direction of airflow. The stream of air after the conversion, which takes place in nozzles/stationary guide vanes, possesses a very high velocity head. The impulse force produced when the high velocity stream of air hits on the blade is utilized for the rotation of the rotor. Unlike a reaction turbine, there will not be any pressure variation in the stream of air as it moves through the rotor blades. Therefore, the rotation is based purely on the impulse force. Since the rotor blades and guide vanes are symmetrical when viewed from both ends, the direction of turbine rotation is same regardless of the airflow direction (Figure 5.12), making it self-rectifying in nature (Setoguchi et al. 2000, 2002). Radial turbines are also superior in terms of the torque obtained from the rotor due to its radial configuration. Despite high damping experienced in the radial turbine, they are favorable due to their durability and low maintenance cost.

5.6 Optimization of Air Turbines

Optimization techniques are frequently used in renewable energy devices to optimize the overall performance. As the performance of any renewable energy system depends on several parameters, an optimization algorithm is required to find out the optimum design that maximizes the performance. In the case of wind turbine design, it is very common practice to use optimization algorithms for turbine blade designs. Chehouri et al. (2015) provide a detailed analysis of optimization methods practical to wind turbine blade design. Banos et al. (2011) explain about the optimization techniques used

Figure 5.12 Flow over an impulse turbine.

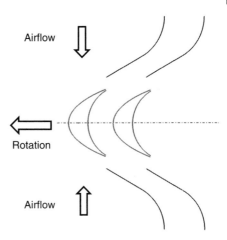

Airflow

Rotation

Airflow

in renewable energy applications. However, there is not much mention about the optimization techniques used in wave energy optimization. In the last decade, different optimization techniques have been employed to optimize wave energy turbines for better efficiency and improved performance characteristics. The optimization techniques used include gradient-based optimization, direct evolutionary optimization, surrogate-based optimization, and automated optimization using genetic algorithms to name a few. Such optimization methods are employed to both impulse and reaction turbines used in wave energy conversion systems.

Gato and Henriques (1996) optimized the symmetrical blade profile of a Wells turbine based on a two-dimensional potential flow calculation and Polak-Ribiere's conjugate gradient algorithm. The aim was to control the pressure distribution shape along the blade profile and delay the separation and stall so as to extend the turbine's operating range. The optimization is carried out using conventional conjugate gradient methods. The design space is defined by a set of geometric parameters describing blade geometry. The blade shape was described using NACA four-digit parameters and the geometric parameters used are trailing edge thickness, trailing edge angle, maximum thickness, and leading edge radius. The objective function in this case is to minimize adverse pressure gradient around the blade. The result of optimization shows improvement in the operating range of the turbine.

Surrogate-based optimization is used to optimize both impulse and reaction turbines. Badhurshah and Samad (2015a) created multi-objective optimization (MOO) based on a genetic algorithm assisted by multiple surrogates on an impulse turbine to increase its performance. The number of stator and rotor blades were taken as the design variables and the objective functions were the pressure drop minimization and turbine shaft power maximization. A weighted average surrogate (WAS), kriging, neural network, and a response surface approximation were used to create a population for the MOO technique and a Pareto-optimal front of the objectives was generated. The surrogate modeling is an approximate technique to use numerical optimization with flow analysis and geometry modification. The flow chart for design optimization is presented in Figure 5.13. Figure 5.14 displays an efficiency comparison for the surrogate and reference models.

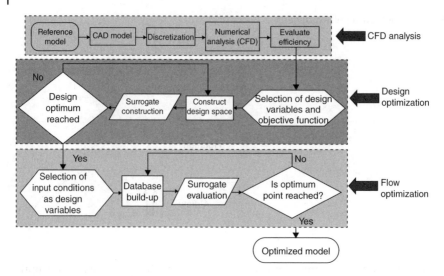

Figure 5.13 Flow chart of surrogate modeling. Source: Reprinted from Badhurshah and Samad 2015b (figures 2 and 6 from original source) with permission from Elsevier.

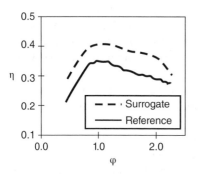

Figure 5.14 Efficiency comparison between surrogate predicted model and reference. Source: Reprinted from Badhurshah and Samad 2015b (figure 6 from original source) with permission from Elsevier.

Gomes et al. (2012) used a two-step optimization method of a two-dimensional blade section of an axial impulse turbine. In the first step, an inverse design method was employed where a constant pressure load was imposed by varying camber line slope alongside the axial chord. In the second step, a thickness distribution optimization compared the distribution of pressure at the trailing and leading edges. Bézier curves were used to outline the thickness of the turbine blade (Figure 5.15). The design variables in this problem were the Bézier curve control points that express the thickness distribution along the blade section. The objective function of this study was to increase turbine efficiency. The result showed improvement in rotor efficiency compared to a standard design.

Optimization techniques are also used in the Wells turbine to improve its performance. Similar to impulse turbines, surrogate modeling techniques are used in the Wells turbine also for the purpose of performance optimization. The different design variables and objective functions chosen are tabulated in Table 5.1.

Mohamed et al. (2011) and Mohamed and Shaaban (2014) used an automated optimization algorithm to optimize the blade profile of a Wells turbine. Automated optimization is a technique where optimization and CFD analysis are carried out within

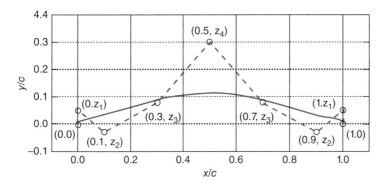

Figure 5.15 Bazier curve with nine control points.

Table 5.1 Design variables for Wells turbine optimization.

Design variables	Objective function
Blade sweep angle at tip and hub (Halder et al. 2017)	Maximize peak torque coefficient and efficiency
Blade sweep angle at tip and mid span and blade thickness at tip and hub (Halder and Samad 2016)	Maximize peak torque coefficient and efficiency
Turbine rotational speed and inlet air velocity (Halder and Samad 2016)	Maximize peak torque coefficient and efficiency

the same iteration loop. Automated optimization is carried out using an optimization library OPAL, written by researchers from Otto von Guirecke University, Magdeburg, Germany. The optimization algorithm uses Gambit for CAD model and grid generation and CFD software ANSYS FLUENT for the computing flow field across the improved Wells turbine. Such methods are automated via journal scripts of Gambit and FLUENT so that human intervention is not required once the process starts. For performing the blade profile optimization 12, several points are considered along with the boundary of the blade profile. The points are given an upper and lower limit (Figure 5.16) so that the optimized blade shape remains feasible for manufacturing. After each iteration, the points take different values and a new blade shape is created followed by a CFD simulation. Such a process continues till an optimized blade shape is achieved. For this study, the initial reference blade is considered as NACA0021 and the modified blade shape found after the optimization is shown in Figure 5.17. The optimized blade shows a 11.3% improvement in power output and efficiency increases by 1%. Using the same optimization technique, optimization of blade pitch angle is carried out for finding out the optimum pitch angle for Wells turbine blades. It is found that the optimized blade pitch angle for the NACA0021 blade is +0.30.

Shaaban (2017) introduced a blade profile optimization technique that improves the air turbine performance while considering the complex 3D flow phenomena. This technique produces non-standard blade profiles from the coordinates of the standard ones. It implements a MOO algorithm in order to define the optimal blade profile. In this case,

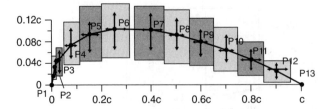

Figure 5.16 Parameters of the blade profile. Source: Reprinted from Mohamed et al. 2011 (figure 7 from original source) with permission from Elsevier.

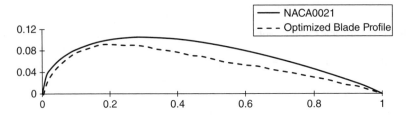

Figure 5.17 Comparison of NACA0021 and optimized blade profile. Source: Reprinted from Mohamed et al. 2011 (figure 10 from original source) with permission from Elsevier.

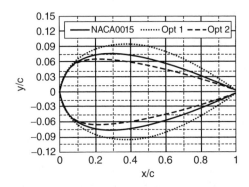

Figure 5.18 Original and optimized blade profiles. Source: Reproduced with permission from Shaaban 2017 (figure 8 from original source) © Wiley-VCH Verlag GmbH & Co. KGaA.

the aerofoil is divided into two parts: from leading edge to point of maximum thickness and maximum thickness to trailing edge. Both the parts are modified to find out the optimized non-standard blade profile. Figure 5.18 shows the original NACA0015 and optimized blade profiles.

It is observed that, in all the cases, the optimized blade profiles show an improvement in performance of the turbines.

References

Antonio, F.D.O. (2010). Wave energy utilization: a review of the technologies. *Renewable and Sustainable Energy Reviews* 14 (3): 899–918.

Badhurshah, R. and Samad, A. (2015a). Multi-objective optimization of a bidirectional impulse turbine. *Proceedings of the Institution of Mechanical Engineers, Part A: Journal of Power and Energy* 229 (6): 584–596. https://doi.org/10.1177/0957650915589271.

Badhurshah, R. and Samad, A. (2015b). Multiple surrogate based optimization of a bidirectional impulse turbine for wave energy conversion. *Renewable Energy* 74: 749–760:https://doi.org/10.1016/j.renene.2014.09.001.

Baños, R., Manzano-Agugliaro, F., Montoya, F.G. et al. (2011). Optimization methods applied to renewable and sustainable energy: a review. *Renewable and Sustainable Energy Reviews*:https://doi.org/10.1016/j.rser.2010.12.008.

Benini, E. and Toffolo, A. (2002). Optimal design of horizontal-axis wind turbines using blade-element theory and evolutionary computation. *Journal of Solar Energy Engineering* 124 (4): 357–363.

Ceballos, S., Rea, J., Robles, E. et al. (2015). Control strategies for combining local energy storage with wells turbine oscillating water column devices. *Renewable Energy* 83: 1097–1109:https://doi.org/10.1016/j.renene.2015.05.030.

Chehouri, A., Younes, R., Ilinca, A., and Perron, J. (2015). Review of performance optimization techniques applied to wind turbines. *Applied Energy* https://doi.org/10.1016/j.apenergy.2014.12.043.

Council, G. W. E. (2016). Global wind report 2010. Online: www.gwec.net/index.php.

Dal Monte, A. (2017). Development of an open source environment for the aero-structural optimization of wind turbines. Ph.D. Thesis, Dipartimento di Ingegneria Industriale, Università di Padova, Italy.

Dal Monte, A., De Betta, S., Castelli, M.R., and Benini, E. (2017). Proposal for a coupled aerodynamic–structural wind turbine blade optimization. *Composite Structures* 159: 144–156.

Dixon, S.L. and Hall, C.A. (2014). *Fluid Mechanics and Thermodynamics of Turbomachinery*. Amsterdam: Butterworth-Heinemann/Elsevier.

Drela, M. (1989). XFOIL: an analysis and design system for low Reynolds number airfoils. In: *Low Reynolds Number Aerodynamics*, 1–12. Berlin, Heidelberg: Springer.

Falcão, A. F. O., and Gato, L. M. C. (2012). Air turbines. In *Comprehensive Renewable Energy* (Vol. 8, pp. 111–149). Oxford: Elsevier. https://doi.org/10.1016/B978-0-08-087872-0.00805-2.

Gato, L.M.C. and Henriques, J.C.C. (1996). Optimization of symmetrical profiles for the Wells turbine rotor blades. In: *ASME Fluids Engineering Division Summer Meeting*, vol. 238, 623–630. https://doi.org/10.13140/2.1.4689.7922. ASME Publications.

Gomes, R.P.F., Henriques, J.C.C., Gato, L.M.C., and Falcão, A.F.O. (2012). Multi-point aerodynamic optimization of the rotor blade sections of an axial-flow impulse air turbine for wave energy conversion. *Energy* 45 (1): 570–580. https://doi.org/10.1016/j.energy.2012.07.042.

Halder, P. and Samad, A. (2016). Optimal wells turbine speeds at different wave conditions. *International Journal of Marine Energy* 16: 133–149. https://doi.org/10.1016/j.ijome.2016.05.008.

Halder, P., Rhee, S.H., and Samad, A. (2017). Numerical optimization of wells turbine for wave energy extraction. *International Journal of Naval Architecture and Ocean Engineering* 9 (1): 11–24. https://doi.org/10.1016/j.ijnaoe.2016.06.008.

Hansen, C. (1993). Aerodynamics of horizontal-axis wind turbines. *Annual Review of Fluid Mechanics* 25: 115–149. https://doi.org/10.1146/annurev.fluid.25.1.115.

Jacobson, R., Meadors, M., Jacobson, E., and Link, H. (2003). Power performance test report for the AOC 15/50 wind turbine, test B. In: *National Wind Technology Center*. Colorado: National Renewable Energy Laboratory.

Lewis, A., Estefen, S., Huckerby, J. et al. (2011). Ocean energy. In: *IPCC Special Report on Renewable Energy Sources and Climate Change Mitigation* (ed. O. Edenhofer, R. Pichs-Madruga, Y. Sokona, et al.). Cambridge and New York: Cambridge University Press.

Lindenburg, C. (2003). Investigation into rotor blade aerodynamics. ECN Report: ECN-C-03-025.

Malcolm, D. J. (2003). Market, cost, and technical analysis of vertical and horizontal axis wind turbines task# 2: VAWT vs. HAWT technology. Lawrence Berkeley National Laboratory.

Mohamed, M.H. and Shaaban, S. (2014). Numerical optimization of axial turbine with self-pitch-controlled blades used for wave energy conversion. *International Journal of Energy Research* 38 (5): 592–601. https://doi.org/10.1002/er.3064.

Mohamed, M.H., Janiga, G., Pap, E., and Thévenin, D. (2011). Multi-objective optimization of the airfoil shape of wells turbine used for wave energy conversion. *Energy* 36 (1): 438–446:https://doi.org/10.1016/j.energy.2010.10.021.

Raghunathan, S. and Abtan, C. (1983). Aerodynamic performance of a wells air turbine. *Journal of Energy* 7 (3): 226–230:https://doi.org/10.2514/3.48075.

Raghunathan, S. and Tan, C. (1982). Performance of the wells turbine at starting. *Journal of Energy* 6 (6): 430–431:https://doi.org/10.2514/3.48058.

Sant, T. (2007). Improving BEM-based aerodynamic models in wind turbine design codes. Doctoral thesis. TU Delft. Doi:uuid:4d0e894c-d0ad-4983-9fa3-505a8c6869f1.

Selig, M.S. and Coverstone-Carroll, V.L. (1996). Application of a genetic algorithm to wind turbine design. *Journal of Energy Resources Technology* 118 (1): 22–28.

Setoguchi, T., Takao, M., Kinoue, Y. et al. (2000). Study on an impulse turbine for wave energy conversion. *International Journal of Offshore and Polar Engineering* 10 (2): 145–152.

Setoguchi, T., Santhakumar, S., Takao, M. et al. (2002). A performance study of a radial turbine for wave energy conversion. *Proceedings of the Institution of Mechanical Engineers, Part A: Journal of Power and Energy* 216 (1): 15–22. https://doi.org/10.1243/095765002760024917.

Shaaban, S. (2017). Wells turbine blade profile optimization for better wave energy capture. *International Journal of Energy Research* 41 (12): 1767–1780. https://doi.org/10.1002/er.3745.

Shehata, A.S., Xiao, Q., Saqr, K.M., and Alexander, D. (2017). Wells turbine for wave energy conversion: a review. *International Journal of Energy Research* 41 (1): 6–38.

Soltanmohamadi, R. and Lakzian, E. (2015). Improved design of wells turbine for wave energy conversion using entropy generation. *Meccanica* 51 (8): 1713–1722:https://doi.org/10.1007/s11012-015-0330-x.

Spera, D.A. (1994). Introduction to modern wind turbines. In: *Wind Turbine Technology: Fundamental Concepts of Wind Turbine Engineering*, 47–72. New York: ASME.

"The DeepWind Project 2016." http://www.deepwind.eu/the-Deepwind-Project.

Torresi, M., Camporeale, S.M., and Pascazio, G. (2009). Detailed CFD analysis of the steady flow in a wells turbine under incipient and deep stall conditions. *Journal of Fluids Engineering* 131 (7): 71103. https://doi.org/10.1115/1.3155921.

Nomenclature

a	effective passage width
A	area
$\boldsymbol{A}$	matrix obtained from input values of the DOE
B	blade metal angle, blockage coefficient
BB	blade loading parameter
b	blade height
C	absolute velocity, blade chord length
$\boldsymbol{C}$	identity matrix at the beginning of the search iterations
C_d	energy dissipation coefficient
C_L	lift coefficient
C_p	power coefficient
C_f	skin friction coefficient
COV	covariance matrix
c_f	shear stress coefficient
D	diameter
d	diameter
DR	Lieblein diffusion factor
E	specific hydraulic energy, efficiency
F	off-design flow-speed ratio, force
F_c	combined objective
f	objective function
f_c	circumferential body force
f_r	radial body force
g	gravitational acceleration
H	head, blade height, turbine head, shape factor
h	head, annulus height, adiabatic efficiency, enthalpy
h_c	concentrated pump loss
h_d	distributed pump loss
I	rothalpy, moment of inertia
i	incidence angle
K	experimental-based loss factor, compressibility coefficient
K_c	normalized velocity
K_f	friction coefficient
K_M	off-design coefficient
K_u	normalized circumferential velocity

Design Optimization of Fluid Machinery: Applying Computational Fluid Dynamics and Numerical Optimization, First Edition. Kwang-Yong Kim, Abdus Samad, and Ernesto Benini.
© 2019 John Wiley & Sons Singapore Pte. Ltd. Published 2019 by John Wiley & Sons Singapore Pte. Ltd.

K_δ	empirically derived coefficient
k	specific speed, turbulent kinetic energy
L	diffusion loading coefficient, specific turbine stage work
l	profile chord
l_a	blade axial chord
M	Mach number
M_b	bending moment
m	maximum camber
$(m)_{\sigma=1}$	parameter referring to a sample cascade with solidity equal to 1
$\dot{m}$	mass flow rate
N	rotational speed, number of blades
n	number of sampling points
N_{ss}	suction specific speed
P	pressure, fan power, power
p	order, pressure
p^0	total pressure
P_R	fan power
Q	volumetric flow rate
Q_f	leakage flow rate
R	radius, correlation coefficient, Pearson's correlation coefficient, reaction of degree
R_B	radius of the bubble
R_b	radius of curvature of blade
r_c	radius of curvature of streamline
S	entropy, source term
s	profile thickness
T	temperature, angular pitch
t	maximum thickness, Student t parameter
Th	blade LE thickness
U	blade velocity, circumferential velocity, circumferential blade speed
u	axial velocity
V	absolute velocity, average velocity at fan outlet
v	tangential velocity
W	blade tangential velocity, relative velocity, blade width
$\mathbf{W}$	diagonal matrix of weight coefficients
w	radial velocity
w_f	weighting factor
$\mathbf{X}$	vector of decision variables
x	independent variable
Y	Ainley loss coefficient
$\mathbf{Y}$	vector of associated fitness functions
y_{CL}	control point
$\mathbf{y}$	vector of output responses from DOE
$y+$	normalized height
Z	number of blades, elevation of a point
x, y, z	Cartesian coordinates
r, θ, z	cylindrical coordinates

m', θ, z	conformal coordinates
$\hat{e}_n$, $\hat{e}_\theta$, $\hat{e}_m$	unit vectors in cylindrical coordinates

Abbreviations

AEP	annual energy production
AMCA	air movement and control association
AMGA	archive-based micro-genetic algorithm
ANN	artificial neural network
AO	automated optimization
AOD	arbitrary optimal design
BEP	best efficiency point
BEM	blade element method
BHP	brake horsepower
BOS	balance of station
CCD	charge-coupled device
CFD	computational fluid dynamics
COE	cost per unit of energy
DCA	double circular arc
DFR	downstream flow resistance
DNS	direct numerical simulation
DoE	design-of-experiment
DPIV	digital particle image velocimetry
EG	error goal
FCR	fixed charge rate
FEA	finite element analysis
FEASM	finite element approximate solution method
GA	genetic algorithm
GMDH	group method of data handling
PRESS	predicted error sum of squares
HAWT	horizontal axis wind turbine
HVAC	heating, ventilation and air-conditioning
ISO	International Organization for Standardization
IGV	inlet guide vane
LOS	language for OPAL++ scripting
KRG	kriging surrogate model
LB	lower bound
LE	leading edge
LDA	laser doppler anemometry
LDV	laser doppler velocimetry
LES	large eddy simulation
LHM	Latin hypercube method
LHS	Latin hypercube sampling
MDO	multi-disciplinary design optimization
MGV	misaligned guide vane

MI	matching index
MOEA	multi-objective evolutionary algorithm
MOGA	multi-objective genetic algorithm
MoI	moment of inertia
MOO	multi-objective optimization
MORDE	multi-objective robust design exploration
MPI	mess passing interface
MRF	multiple reference frame
MSE	mean square error
NOP	number of operating conditions
NPSH	net positive suction head
NPSE	net positive suction specific energy
NSGA	non-dominated sorting of genetic algorithm
NURBS	non-uniform rational B-spline
ORC	organic Rankine cycles
OPAL	optimization algorithm library
OWC	oscillating water column
PADRAM	parametric design and rapid meshing
PBA	PRESS based averaging
PDA	phase doppler anemometer
PIV	particle image velocimetry
POSs	Pareto-optimal solutions
PR	pressure ratio
PRESTO	pressure staggering option
PRESS	predicted error sum of squares
PS	pressure side
RANS	Reynolds-averaged Navier–Stokes
RE	radial equilibrium
RBNN	radial bias neural network surrogate model
RSA	response surface approximation surrogate model
RNG	re-normalization group
SC	spread constant
SLC	streamline curvature
SM	stall margin
SOPHY	soft-PADRAM-hydra
SPL	sound pressure level
SQP	sequential quadratic programming
SGS	sub-grid scale
SS	suction side
SST	shear stress transport
TC	turbine cost
TEIS	two element in series
UB	upper bound
URANS	unsteady Reynolds-averaged Navier–Stokes
VAWT	vertical axis wind turbine
VSV	variable stator vane
WAS	weighted average surrogate

Greeks

Ω_{ij}	rate of rotation tensor
$\hat{\beta}$	estimate of coefficient vector b in polynomial regression
$\emptyset$	viscous dissipation, scalar quantities
α	absolute flow angle, step length parameter, angle of attack, weight factor
β	relative flow angle, vector of regression coefficients
β_i ; β_{ij}	coefficients of the basis functions in polynomial regression
Γ	diffusivity
γ	sweep angle, surface tension, stagger angle, twist distribution
δ^*	boundary layer displacement thickness
δ	deviation angle, tip clearance gap
ε	slip factor, total blockage coefficient, deflection, lean angle, dissipation rate of turbulent kinetic energy
ε	vector of random errors
ζ	efficiency
ζ_s	secondary loss coefficient
ζ_{sh}	shock correction
ζ_δ	tip clearance loss coefficient
η	efficiency
θ	angle between inlet and outlet ports, profile camber, momentum thickness of boundary layer, coning angle
λ	average of the inlet and outlet areas of the impeller
μ	mean, slip factor, dynamic viscosity
ρ	density
σ	slip factor, standard deviation, cavitation coefficient, Thoma's cavitation factor, solidity, normal stress
τ	radial tip clearance, stress
υ	kinematic viscosity
φ	discharge coefficient
ϕ	flow rate, potential function, viscous dissipation, scalar quantities
χ_M	Mach-based correction
χ_R	Reynolds-based correction
ω	total pressure loss coefficient, angular velocity
$(\delta_0)_{10}$	basic deviation for symmetrical 10% thickness-to-chord profile
Θ	momentum thickness of boundary layer
χ	margin of safety
δ	deviation angle, tip clearance gap, tilt angle
ν	hub to tip ratio

Subscripts

0	stagnation property, inlet duct
1	inlet of impeller, upstream

2	outlet of impeller, downstream
3	vaneless diffuser exit
a	atmospheric
ad	adiabatic
at	atmospheric
B	blade, bubble
b	blade
back	backward
c	critical, blade metal, centrifugal
d	disk
e	edge quantities at the boundary layer thickness
F	front, fluid
for	forward
h	hydraulic
HUB	hub
hyd	hydraulic
i	inlet, initial
in	inlet
im	impeller
M	meridional
m	meridional, operating condition of model, midspan
max	maximum
mech	mechanical
min	minimum
n	number of decision variables, normal
o	outlet
out	outlet
p	operating condition of prototype, profile
R	required, rotor
r	radial
s	specific speed
ref	reference
req	required
rms	root mean square
S	static, suction, stator
s	static, suction, search direction, secondary
T	total
TIP	tip
t	tip, total, target, turbine, turbulent
ti	tip clearance
ts	total-to-static
U	tangential component
u	circumferential, tangential
v	vapor
vol	volumetric
x	axial
θ	circumferential

Superscripts

.	time rate of change
is	isentropic state
-	average
*	nominal condition
‘	actual angle

Index

Design Optimization of Fluid Machinery: Applying Computational Fluid Dynamics and Numerical Optimization,
First Edition. Kwang-Yong Kim, Abdus Samad, and Ernesto Benini.
© 2019 John Wiley & Sons Singapore Pte. Ltd. Published 2019 by John Wiley & Sons Singapore Pte. Ltd.